ACT Math in the Classroom

Bring ACT Math and Science prep into the classroom to enhance student learning! In this new copublication from Routledge and test-prep experts A-List Education, you'll learn how the updated ACT exam is closely aligned with the Common Core, making it easy to weave test prep into your curriculum and help students hone the skills they need for college readiness. The book is filled with practical examples of how the Common Core State Standards are connected to specific sections, question types, and strategies applicable to the ACT, so you can simultaneously prepare your students for the test while improving their overall math, science, and reasoning skills.

Bonus: A Study Guide to help you use the book for school-wide professional development is available as a free eResource download from our website: www.routledge.com/9781138692213.

A-List Education is an educational services provider serving more than 50,000 students with tutoring programs across the U.S. as well as in the UK, Dubai, Switzerland, and China.

Other Books Available From Routledge and A-List Education

(www.routledge.com/eyeoneducation)

ACT ELA in the Classroom:
Integrating Assessments, Standards, and Instruction

SAT ELA in the Classroom:
Integrating Assessments, Standards, and Instruction

SAT Math in the Classroom:
Integrating Assessments, Standards, and Instruction

ACT Math in the Classroom

Integrating Assessments, Standards, and Instruction

A-List Education

Routledge
Taylor & Francis Group
NEW YORK AND LONDON

A LIST

First published 2017
by Routledge
711 Third Avenue, New York, NY 10017

and by Routledge
2 Park Square, Milton Park, Abingdon, Oxon, OX14 4RN

Routledge is an imprint of the Taylor & Francis Group,
an informa business

Library of Congress Cataloging-in-Publication Data
Names: A-List Education.
Title: ACT math in the classroom : integrating assessments, standards, and instruction / by A-List Education.
Description: New York : Routledge, 2017. | Series: A-List SAT and ACT series
Identifiers: LCCN 2016011085 | ISBN 9781138692206 (hardback) | ISBN 9781138692213 (pbk.) | ISBN 9781315532851 (e-book)
Subjects: LCSH: Mathematics—Examinations, questions, etc. | Mathematics—Study and teaching. | ACT Assessment.
Classification: LCC QA43 .A43 2017 | DDC 510.71/2—dc23
LC record available at https://lccn.loc.gov/2016011085

ISBN: 978-1-138-69220-6 (hbk)
ISBN: 978-1-138-69221-3 (pbk)
ISBN: 978-1-315-53285-1 (ebk)

Typeset in Palatino
by Apex CoVantage, LLC

Contents

eResources

This book is accompanied by free online eResources, including a Study Guide to help you work on this book with colleagues, as well as additional materials to help you with school-wide implementation of the ideas in this book. To access the eResources, go to www.routledge.com/9781138692213 and click on the eResources tab. Then click on the items you'd like to view. They will begin downloading to your computer.

About the Author

A-List Education was founded in 2005 with a mission to bring innovation and opportunity to education, empowering students to reach their true potential. We work with schools, school districts, families, and nonprofits and provide tailored solutions for specific learning and curriculum needs—ultimately working to improve college readiness and access. Our staff comprises experienced and passionate educators each with a distinctive and personal approach to academic success, and our management team collectively possesses 75+ years of tutoring, teaching and test preparation experience. We now provide leading-edge education services and products to more than 500 high schools and nonprofit organizations, helping more than 70,000 students a year in the United States and around the world.

A-List has a variety of offerings for SAT and ACT preparation, including:

- **Textbooks** for students studying individually or for teachers conducting classes. Our content not only emphasizes test-taking techniques but also reinforces core skills, which empower students for academic success long after taking the test.
- **Professional development** to help schools and organizations set up their own courses. Our seminars create valuable educational expertise that will allow teachers in your district to bring content and problem solving strategies directly into their classrooms.
- **Direct course instruction** using our own staff. Our dedicated and experienced teachers receive intensive training before entering the classroom, and have proven track records of empowering students to reach their academic potential.
- An **online portal** to remotely grade practice tests and provide supplemental material. This platform removes the burden of grading complex tests without requiring

customized technology and provides supplemental material for your ongoing courses.

- Individual one-on-one **tutoring**. Our instructors help students deliver average improvements of more than three times the national average in the United States. In addition, our students routinely gain acceptance to their top choice schools and have been awarded millions of dollars in scholarships.

Visit us at **www.alisteducation.com** to learn more.

Introduction and Overview

The ACT and the Common Core

The Common Core State Standards Initiative (CCSSI) is a program designed to unify the state standards across the United States so that students, parents, and educators have a clear sense of what skills students must acquire in school to become ready for college or a career. It is an independent initiative in which states may voluntarily choose to participate, but by the end of 2015, 42 states plus the District of Columbia had chosen to adopt the Standards. The Standards are designed to be rigorous, clear, and consistent, and they are based on real evidence to align with the knowledge and skill necessary for life beyond high school.

The program is controversial to be sure, but a strong foundation is already in place, and schools around the country are working to align their own materials and programs with the newly adopted standards. The Standards specifically do not outline exact requirements for curriculum, such that schools and districts can still make their own choices about how to run their classes while still adhering to the Standards. As a result, some schools may struggle to find the right material.

However, there already exists a comprehensive source of material that addresses the wide range of skills and knowledge that the Common Core emphasizes: the ACT.

Not all schools currently offer preparation programs for the ACT, but even those that do tend to treat the test as distinct from normal schoolwork. The test is seen as supplementary, and preparation is an unpleasant game in which students learn tricks to game the system without actually learning skills. This view of the test, however, is not just uncharitable but false. In fact, ACT preparation can fill many of the gaps to help schools align their curricula with the Common Core.

The ACT requires students to use many of the same math and reading skills that are the goals of the Standards. It is designed to identify whether students are ready for college by testing them on the skills and knowledge they will need when they get there.

Preparing for the test can accomplish two goals at once. Test preparation's primary goal is to prepare for the test itself, helping students maximize their scores on the test and thus improving their chances of being admitted to the colleges of their choice. Beyond pure admissions, students' test scores can have a number of uses for different programs and institutions. For example,

- **College admissions**. Roughly half of a student's admissions profile is composed of a combination of GPA and SAT or ACT score. A high score can be a huge differentiator for the majority of elite universities, and a minimum hurdle for the majority of state universities.
- **Scholarships**. There are billions of dollars of aid in private and school based scholarship money tied directly to test scores.
- **Community colleges**. Even at community colleges with low graduation rates, good scores can allow students to avoid placement in remedial classes.
- **Military**. For students interested in the military, baseline scores can qualify a student for officer training as opposed to regular enlistment.
- **State assessment**. The ACT is increasingly used as a statewide student assessment to identify achievement of particular benchmarks.

But test preparation is also a valuable activity in itself; students will also be working on honing and strengthening the skills they need for college readiness. Furthermore, the test material itself is valuable even beyond their application to the actual test. The passages, mathematical concepts, and other content contained herein can be divorced from the ACT. You do not have to actually take the ACT to draw value from reading and analyzing its passages, or from analyzing its grammatical structures, or from attempting to solve its math problems.

It is for these reasons that some states have decided to use the ACT as its primary measurement for high school achievement, rather than a more explicitly standards-based assessment.

Alignment

In 2010, ACT, Inc., produced an alignment study to show how the standards align with the skills that the ACT assesses. The results of this study show that the ACT significantly aligns with the standards. The ACT is constructed not according to the CCSS but using its own College and Career Readiness Standards (CCRS). The aforementioned study aligns the CCSS to these ACT standards, not to the ACT test itself. Our purpose here is to examine this alignment in the context of real test material. It's fine to say that CCSS X aligns with ACT standard Y, but what does that mean in terms of actual test questions, the things students will actually engage with?

Since the publication of this study, the ACT has altered its ACT CCRS but has not updated the CCSS alignment document. Most of the changes, however, should not affect alignment to CCSS. Many of the changes are cosmetic—for example, changing the wording of existing standards or re-categorizing mathematical concepts—and the test itself has not changed significantly, with the exception of the Writing section's optional essay. The essay prompt has changed format and the scoring system has changed. If anything, these changes associate the ACT CCRS more closely to the CCSS.

We use the ACT's own alignment document as a starting point, and elaborate based on our own extensive knowledge of and experience with test material. Interestingly, this document is no longer available on the ACT's website, possibly because the changes to their ACT CCRS make it somewhat obsolete. It is still possible to find it online hosted elsewhere with a bit of searching.

In short, the ACT was already very strongly aligned to the CCSS before they changed their standards. Since then, the test has changed very little and their standards have moved *closer* to the CCSS.

About This Book

This book has two main goals:

1. To show how specific ACT Math material aligns with the Math CCSS.

2. To discuss how to incorporate ACT preparation into your regular math classes outside of an explicit test preparation class.

Why do we focus on the CCSS? First of all, because the grand majority of states use it. There's no shortage of debate about the value of the standards, but it's undeniable that they are in place throughout most of the country. Even some states that are moving away from the standards are doing so more in name than in practice and are keeping the content of the CCSS in place.

Of course, not every state uses CCSS, but it remains a useful framework for connecting ACT material to classroom material. We can't run through every state's particulars in one book, but the popularity of the CCSS makes them a convenient reference. If you don't use them, you can still use CCSS as a touchstone to compare to your own state standards. Even if you do use CCSS, your state may also have its own assessments or graduation requirements that deviate from CCSS. The alignment information is one piece of the picture.

Second, some states are actually using the ACT as their main statewide assessment. They are generally doing so because of the ACT's alignment to the Common Core. Is this a good idea? That's debatable, and we won't take sides here. The question is part political, part pedagogical, often emotional. However, we can offer our research and expertise in the test so you can help see for yourself where it coincides with your curriculum and where it doesn't.

Structure of the Book

Part 1 will describe the structure and content of the ACT's Math sections for those who may be unfamiliar with the test, along with discussion of how the test has changed and some effective test-taking techniques. Keep in mind that this is a general overview. It draws information from our main textbook, *The Book of Knowledge*, which is the product of years of experience with the test and goes into much greater detail about the test's content and the most effective strategies.

Part 2 will connect ACT Math to the Common Core Math standards. This section will take a closer look to show where the test does and does not align with these Standards. It will go through each content category and domain to discuss what specific sections, question types, or strategies align with the standards in question. Additional discussion also describes how, even when the test does not align with the standard, test material can be pushed beyond its intended scope in order to do so. Additionally, we will discuss how ACT Science aligns to the CCSS, which will stray past the math standards into the reading standards.

Parts 3 and 4 will discuss how to reconcile this material with your classroom. This could mean using your math classes as explicit preparation for the test; it could mean using test material to supplement your regular classes; it could mean preparing for the test as a tool with which to get your students to meet the standards; or it could be a combination of these things.

The Appendices will list all of the complete alignment tables discussed in Part 2 as well as a bibliography and suggestions for further reading.

1

About the ACT

In recent years, the ACT has made an active effort to emphasize college readiness benchmarks to a greater degree than they did before. They have introduced ACT Aspire, a suite of assessment for students starting as early as third grade, using a longitudinal assessment of benchmarks into high school. They are moving past the market of individual students taking the test for college and marketing their tests directly to schools, to be used both for college admissions and for skills assessment.

Is this a good idea? Who knows? There's certainly a great deal of debate about the value of assessments in general, let alone this particular test. But in the meantime, it's here, so we must acknowledge it and deal with it. The test has been relatively unchanged for several decades (with a few exceptions we'll discuss soon), so we do know quite a bit about what's on the test and have years of experience preparing students for it.

We want to help you incorporate ACT material into your classroom in order to prepare students without running an explicit ACT prep course. To do so, the first and most important thing is simply to know what's on the test and what the test is like. The best way to do that is firsthand: *You should absolutely do some official practice tests yourself.* There is a full-length practice test available for free download on the ACT's website. Go do one and see what you think.

In the meantime, we're not going to go through our whole prep book (but it is for sale on our website!), but we do want to give you an overview of the structure and content of the test.

Format

The ACT is composed of four sections (called "tests"), plus an optional fifth, always presented in this order:

Table 1.1 ACT Format

Test	Number of Questions	Time	Description
English	75	45 minutes	Multiple-choice grammar and usage questions.
Math	60	60 minutes	Multiple-choice math questions.
Reading	40	35 minutes	Multiple-choice reading comprehension questions.
Science	40	35 minutes	Multiple-choice data interpretation questions.
Writing	*1 essay*	*40 minutes*	*Optional. One 1- to 4-page essay.*
Total	**215**	**2 hours 55 minutes**	
with essay	*215 + essay*	*3 hours 35 minutes*	

Most multiple-choice questions have four choices, except for questions on the Math Test, which have five choices. The letters of the answer choices alternate ABCD/FGHJ every other question. On the Math Test, the five choices alternate ABCDE/FGHJK.

The ACT does not take off points for wrong answers. This means that random guessing will not count against you. A wrong answer counts the same as a blank.

Students will get a score from 1 to 36 on each of the four main tests, plus a Composite score, which is the *average* of the four test scores—English, Math, Reading, and Science—rounded to the nearest whole number.

Everyone will also get a STEM score, which is just the rounded average of your Math and Science scores. This score will *not* be included in your Composite score.

Students who choose to take the Writing section (the essay) will get additional scores:

- Four subscores from 2 to 12 in different aspects of your essay

- A final Writing score from 2 to 12 that is the rounded average of the four subscores. This is a change, starting with the September 2016 test. In the 2015–2016 school year, students receive a scaled writing score on the 1 to 36 point scale.

When the Writing score was a scaled 36-point score, students who took Writing also received an ELA score on the 36-point scale, which was the rounded average of the English, Reading, and Writing scores. While the Writing score will no longer be on that scale, the ACT said the ELA score will continue to be reported. However, as of this printing, the ACT has not announced how the ELA score will be calculated.

ACT scores are calculated by taking the number of right answers (the "raw score"), and translating that score into a final score using a special scoring table. Each test has its own unique scoring table in order to adjust for slight difficulty differences among tests. The national average Composite score is generally around 21, but average scores can vary by state and by component test.

For a detailed look at test format, ACT, Inc., publishes a book called *The Official ACT Prep Guide 2016–2017*, which contains three official full-length tests.

Recent Changes

The ACT has not undergone any radical changes like the SAT has. However, in the past few years, there have been a few changes that may be noticeable.

- The essay format and scoring system has radically changed as of September 2015. The old essay featured a different kind of prompt and did not use the 36-point scale (students only got one score from 2 to 12). The STEM and ELA scores are also new additions.
- Starting in 2014, one of the four passages on the Reading Test will now be a double passage, with two different passages on a similar topic and one set of questions about them.
- The distribution of passage types has changed slightly on the Science Tests. Previously the test always contained seven passages, each with five to seven questions, following strict rules about number of questions per passages. As of 2015, some tests have had only six passages with more fluidity in the number of questions per passage, and future tests may have six or seven passages.
- The Math Test had a similar change in the distribution of question types. However, the change in Math is so subtle that no one would notice it unless he or she were looking closely.

If you are using practice tests published before 2015, such as the older book, *The Real ACT Prep Guide 3rd Edition*, the tests will *not* reflect these changes. For the most part that's fine; a seven-passage Science Test isn't significantly different from a six-passage Science Test. The essay change is significant—you shouldn't waste time doing essays in the old format—but the essay is optional, so you can always skip it entirely if doing an older test. Their most recent book, *The Official ACT Prep Guide 2016–2017*, does contain new-format tests, as does the current test that's available for free download on actstudent.org (form 1572CPRE).

This book will focus on the ACT Math and Science Tests. We will not be discussing the English, Reading, or Writing sections.

Math

Format and Content

The Math Test is composed of one section with 60 questions. The questions are ordered by difficulty such that the early questions are easy and the later questions are difficult. All questions are multiple choice, but this is the only section on the ACT in which the questions have five choices instead of four. Calculators are allowed for the Math Test, but not required.

Questions will cover these content areas:

- **Pre-Algebra** (20–25%). These questions include any question that deals with only numbers (not variables), word problems that don't require writing equations, and simple equations with only one variable. Major topics include solving one-variable equations, manipulating fractions and decimals, basic roots and exponents, ratios and percents, statistics and data interpretation.
- **Elementary Algebra** (15–20%). These questions involve more advanced algebraic manipulation. Major topics include multi-variable equations, linear equations, translating word problems into algebraic expressions, substitution of expressions, polynomials, and factoring expressions.
- **Intermediate Algebra** (15–20%). These questions involve higher-level algebra questions and more advanced topics. Major topics include: algebraic expressions using exponents and exponent rules, functions and $f(x)$ format, quadratic manipulation, absolute value, arithmetic and geometric sequences, imaginary numbers, matrices, and logarithms.
- **Coordinate Geometry** (15–20%) These questions include anything and everything related to graphing and the xy-coordinate plane. Major topics include: slope and intercepts, graphing linear equations, circles, ellipses and inequalities, intersection of graphs, midpoint and distance formulas, and transformation of functions.

- **Plane Geometry** (20–25%). Plane Geometry questions include anything and everything related to geometry and figures, (excluding figures graphed on the coordinate plane, which are classified as Coordinate Geometry). Major topics include: angle rules and properties, properties of triangles, rectangles, parallelograms, trapezoids, circles, and calculating area, perimeter, surface area, and volume.
- **Trigonometry** (5–10%). Most trigonometry questions require a basic understanding of trigonometric terms, though some questions may involve higher-level formulas. Major topics include: definitions of basic trigonometric ratios, basic identities, graphing trigonometric functions and their transformations, law of sines and law of cosines, and the Unit Circle.

ACT Math Techniques

While much of the test comes down to straightforward knowledge of math, the test isn't exactly like the kind of tests students see in school. A-List's ACT book, *The Book of Knowledge*, has a number of tips and strategies for the special kinds of issues that arise on the ACT.

Error Avoidance

One of the biggest obstacles students face is *distractor choices*. For the most part, the wrong answers on the question aren't random. They're carefully selected to be tempting for students. There are specific reasons why a student may reasonably believe them to be correct. It could be because of a careless math mistake, like not distributing across parentheses or adding instead of subtracting. It could be because they didn't read the question carefully, like solving for x when the question asks for y.

Techniques

In general, ACT math is not about knowing a lot of content as much it is about knowing a limited amount of content *very well*. The number of rules students need is small compared with what you actually teach in school, but these rules will be combined in unexpected ways. Therefore, learning multiple methods of problem solving is a crucial part of preparing for the test.

For example, often a question will appear to require complex algebra, but the problem can be solved with simple arithmetic. This can be done by choosing values for the variables (we call this *Plug In*) or by testing the values in the choices (we call this *Backsolve*). It takes practice to get comfortable using methods beyond what you're used to, but these methods are crucial to score improvement. If your students don't change their methods, they won't change their scores.

Science

Format

The Science Test is composed of 40 questions split across six or seven passages, each followed by five to eight questions. Each passage consists of a description of a set of situations, experiments, or hypotheses, usually accompanied by some set of tables, graphs, or other figures.

Each passage will focus on concepts taken from one of four major fields of science taught in school: Biology, Chemistry, Earth Science, and Physics. There will be at least one and no more than two passages from each field.

There are three types of passages on the Science Test:

- **Data Representation** (12–16 questions, 30–40% of the test). These will present a set of data describing a situation or phenomenon and ask students to interpret the figures provided. Questions about these passages are generally straightforward and simply require you to read and interpret the information presented.
- **Research Summary** (18–22 questions, 45–55% of the test). These will present a series of experiments on a common theme or topic. In addition to interpreting data, these questions will ask about the design of the experiment or conclusions that can be drawn from the data.
- **Conflicting Viewpoint** (6–8 questions, 15–20% of the test). This will present a situation or occurrence, followed by two or more conflicting hypotheses that seek to explain it. Questions will ask students to compare the different viewpoints.

Strategies

Students often associate the Science Test with the Math Test, since math and science are so closely related. But in terms of the way to do questions, the ACT Science Test is entirely different from the Math Test. There will be numbers, tables, and graphs, but students won't have to *do math*. Calculators are neither allowed nor necessary on the Science Test. While there might occasionally be some light addition or subtraction, usually the most that students will have to do with all these numbers is to judge relative quantity or relationships.

In fact, the Science Test is a lot more like the Reading Test. Like reading passages, science passages will present some information and ask questions about the information presented. And like the Reading Test, questions can be answered simply by *going back to the passage* and looking up the necessary information.

In fact, for the grand majority of questions on the Science Test, *students do not need to know scientific facts* other than what's actually presented in the passage. Questions that explicitly ask for scientific knowledge that is not given in the passage or figures occur very rarely. Slightly more common are questions that require students to know or understand the definition of a concept (like "density") in order to interpret the given data. All these together make up 5% of all questions. That means 95% of the questions can be done purely based on the information given to you on the page.

Question Types

A-List's textbook, *The Book of Knowledge*, categorizes ACT Science questions into the following question types based on our extensive research into the test:

- **Data Lookup Questions**. These are the most common type of question on the Science Test. They require students to retrieve a value given in one of the tables or graphs provided. Often, students simply must match the name of the quantity that the question asks for with the headings and labels in the figure.
- **Combination Questions**. These questions require students to pull information from more than one part of the passage. These questions are often similar to Data Lookup Questions, which require using multiple tables or graphs to find the needed data. Questions that compare experiments, hypotheses, or viewpoints also fall into this category.
- **Relationship Questions**. Instead of asking to retrieve specific data values, these questions ask about the relationship between fields. Usually this is just a matter

of figuring out up vs. down: when one value goes up, does the other value go up, go down, do both, or stay the same?

- **Inferred Data Questions**. These questions are similar to Data Lookup Questions with one important twist: the data point that the question is asking for does not literally appear in the figure. Students must infer the value based on the values that do appear. The point will either be right in between points that are given, or it will be higher or lower than the limits shown.
- **Passage Questions**. These questions require reading the introductory material in the passage and the definitions of terms used in addition to looking at the figures and tables. This could mean simply finding the definition of a term or having a deeper understanding of a concept.
- **Experiment Design Questions.** These questions ask about the *design* of an experiment. These questions will *only* appear in Research Summaries passages, which feature several different experiments around the same topic or line of investigation. They may ask how an experiment was literally set up, why things were set up a certain way, what factors were varied, or what might happen if the experiment is changed.
- **Reasoning Questions**. These questions ask students to make conclusions about certain hypotheses. They may ask whether a certain conclusion is justified, whether a certain fact weakens or strengthens a hypothesis, or what assumptions are necessary to make an argument. Sometimes these questions may require extended reasoning, but often they require little more than looking up data and can be done purely by elimination.
- **Knowledge Questions**. These questions require additional scientific knowledge that is not given in the passage. This may be something as simple as knowing what a word means (like "mammal"), or understanding certain properties or concepts (like "density") that aren't explicitly defined in the passage. These questions can be the most difficult, but they are also the least common question type, occurring only 5–10% of the time.

Conflicting Viewpoint Passages

Data Representation and Research Summary passages tend to have similar questions following them. However, the Conflicting Viewpoint passages are noticeably different. The passage will present a topic or situation followed by several different theories that seek to explain the situation. They usually involve more reading and less data and figures. Conflicting Viewpoint passages frequently feature the following question types we've already seen, but with a focus on the relationship between the different viewpoints:

- **Passage Questions**: Questions that literally ask what the introduction or one of the hypotheses says.
- **Combination Questions**: Questions that ask students to directly compare and contrast the hypotheses.
- **Reasoning Questions**: Questions that ask whether a piece of information would strengthen or weaken one of the theories.

ACT vs. SAT

With the recent redesign to the SAT beginning in 2016, many users of the old SAT have been looking to move to the ACT instead. The redesigned SAT is actually much more similar to the ACT than the old SAT was. Let's take a quick look at the differences between the tests.

Table 1.2 Differences between ACT and SAT

Section	ACT	SAT	Content Differences
English (called "Writing" on SAT)	75 questions 45 minutes 0.6 min/q	44 questions 35 minutes 0.8 min/q	Content between the two tests is virtually identical. SAT has some passages that have accompanying tables or figures.
Math	60 questions 60 minutes 1 min/q	58 questions 80 minutes 1.4 min/q	Math concepts are mostly the same. The SAT: • has a no-calculator section • has two sections • has less geometry, more statistics • has some non-multiple-choice questions
Reading	40 questions 35 minutes 0.88 min/q	52 questions 65 minutes 1.25 min/q	Question types are mostly the same. The SAT: • has passages that come with tables or figures Here is a general overview of the sections: • explicitly asks to find evidence • does not have a Humanities passage • has some older, primary source passages
Science	40 questions 35 min 0.88 min/q	None	No Science section on SAT. Data figures on reading and writing passages involve similar concepts.
Writing (called "Essay" on SAT)	40 minutes	50 minutes	SAT essay prompt is based on reading and analyzing a given passage.

- The biggest difference is in timing. Students have less time per question on the ACT. Students who have particular problems with timing should seriously consider taking the SAT.
- The most noticeable content difference is that the SAT does not have a Science section. However, SAT Reading and Writing sections contain some questions about interpreting tables and graphs in a context, questions that require similar skills as the ACT Science Test.
- The Math sections show the biggest difference in format. The SAT splits its math across two different sections, instead of one big section on the ACT. On one of those sections, calculators are not permitted, while they are permitted on all math questions on the ACT. And each of the two SAT Math sections contain some "Grid-In" Questions for which students must provide their own answers, whereas all ACT questions are multiple choice.
- However, the two Math Tests use very similar math content. The biggest difference is the distribution of concepts. The SAT has much less geometry, less than 10% of the questions on the SAT, and much more statistics and data analysis, about 30%.
- The Reading and Writing content are also very similar between the two tests. The Writing in particular is shockingly identical.
- The essay prompts are in very different formats, but both tests' essays are optional.

2

Alignment with Common Core Math Standards

How to Read the Math Standards

There are two components to the math standards.

First, the *Standards for Mathematical Practice* describe skills necessary for effective mathematical thinking and proficiency. Rather than specific content, these standards refer to habits and practices that students should develop and utilize throughout all their mathematics courses, from kindergarten through twelfth grade.

Second, the *grade-level standards* describe what students should be learning for every grade level they achieve throughout their schooling. Since we are concerned with the ACT, we will focus solely on the standards for high school. Unlike the ELA standards, the mathematics standards are not broken up into individual grades for high school; there is one set of mathematics standards for all of high school.

These standards are divided into several *conceptual categories*:

- Number and Quantity
- Algebra
- Functions
- Geometry

- Statistics and Probability
- Modeling

These roughly correspond with individual courses that students would take in high school, but they do not have to be so discrete. Standards from one category may well also apply to work in courses that focus on other categories.

Each of the standards within the categories is further arranged into different groupings. Standards are arranged into *clusters* of closely related concepts. Closely related clusters are grouped into *domains*. Domains are grouped into conceptual categories. Note that standards are numbered continuously within a domain. Numbering does not restart with a new cluster, only with a new domain. Furthermore, some standards may be subdivided into two or more subskills that are related to that standard.

Each standard has a code containing three parts: a letter denoting the conceptual category, a series of letters denoting the domain, and a number denoting the standard. If the standard has any subskills, the skills will be denoted by a letter following the standard number. For example, "F-IF.7c" denotes:

- the Functions category [F],
- the "Interpreting Functions" domain [IF],
- the seventh standard within that domain [7], and
- the third subskill associated with that standard [c].

Code: [category]-[domain].[standard][subskill].

Modeling is slightly different from the other conceptual categories listed above. The standards in this category have less to do with specific content and more to do with applying math content and skills to real-life situations. Skills like these are integral to the SAT, and questions that ask students to determine how to apply mathematical concepts to actual settings to solve specific problems are prevalent. Rather than producing a separate list of modeling standards, these modeling standards are distributed throughout the other conceptual categories. In all alignment tables, any standard that is also a modeling standard will be marked with an asterisk (*).

Finally, there will be some standards shown here that are more *advanced* than others within their category. The goal of the Common

Core State Standards is to ensure that students are ready for college and future careers, and the standards were specifically written with that in mind. However, within these categories there are also some standards aligning to skills that are not necessary for college and career readiness, but may be necessary for more advanced mathematics courses such as calculus. Standards that go beyond college and career readiness are marked with (+).

Alignment

How Was Alignment Determined?

To determine alignment, we started with the ACT, Inc.'s own 2010 study aligning the ACT Standards to the CCSS. From there, we elaborated in order to contextualize to test content and made adjustments according to how the test has changed.

The following chart outlines the results for math standards, showing what percent of the standards in each strand align with the ACT's stated skill set according to their alignment study.

Table 2.1 Percent of Standards Aligned with ACT Skill Set

Common Core State Standard Category	ACT Alignment
Standards for Mathematical Practice [MP]	**88%**
Standards for Mathematical Content, Grades 9–12	**100%**
Number and Quantity [N]	100%
Algebra [A]	100%
Functions [F]	100%
Geometry [G]	100%
Statistics and Probability [S]	100%
Modeling (*)	100%
Advanced (+)	100%
College and Career Ready (not (+))	100%

As you can see, the ACT considers itself to be almost entirely aligned with the CCSS.

The actual alignments performed were for a larger suite of materials than just the ACT. The ACT, Inc. offers additional tests and programs besides the main college admissions tests. At the time the alignment document was produced, these included EXPLORE and PLAN, given to students earlier in high school (these tests have since been discontinued and replaced with the PreACT). Each test was aligned against the standards for the grade level the particular program is meant to be given. This document will focus on alignment for the ACT Test itself, which was aligned to the Standards for Mathematical Practice and the Standards for Mathematical Content for high school.

Some standards show what we call "partial alignment". In the ACT alignment report, the portion of the standard that is aligned with the ACT was highlighted in yellow. If part but not all of a standard aligned with the ACT, only those parts that align were highlighted. The table shows percentages of standards that show any alignment; it does not distinguish partial and complete alignment.

For more information about how alignment was determined, please refer to the study itself, which is listed in the bibliography at the end of this document.

How to Read This Section

The following discussions will be grouped into chapters by conceptual categories shown above. At the start of each chapter, we will list all of the domains and clusters within that category. For each cluster, the number of standards within that cluster is given in parentheses. The number of subskills associated with those standards is also given, where applicable.

For the sake of clarity, we do not list all individual standards in the discussion chapters. All standards and their alignment to the ACT are shown in the tables in the appendix. Modeling and advanced standards are not treated in separate chapters, but are instead discussed within the context of the relevant subject category.

Alignment

Following the tables, we will briefly summarize the alignment of the ACT with the standards. If a standard is partially aligned—that is, the document considered it aligned, but with a qualification, or it only considered a portion of the standard to be aligned—we will list here which segments do and do not align.

Discussion

Here we discuss why and how the standards align and don't align. For those that don't align well with the test itself, we discuss how the standards might align to the skills and techniques used during

Figure 2.1 Sample Headings

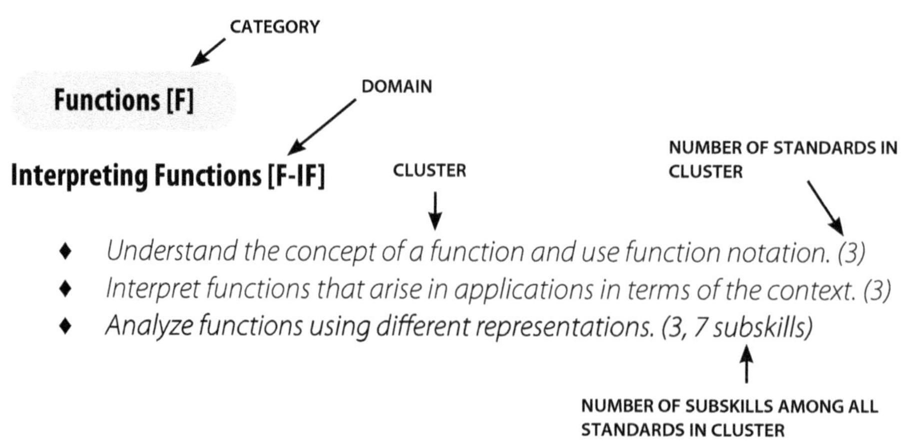

the act of preparation. Furthermore, we discuss ways to incorporate material from the test into tasks beyond the scope of the test. Note the original alignment document often does not give detailed explanations about why a standard does or does not align with the test. Any discussion of such here is our own judgment based on our extensive experience and knowledge of the ACT.

Summary

We give a summary of the previous discussion. This summary may also list suggestions for how to use test material in ways beyond the test's scope in order to meet a standard that otherwise does not align.

Sample ACT Questions

When a domain contains standards that align with the test, we provide some sample questions that demonstrate the concepts from the standards in that domain. All sample questions come from the ACT's book *The Official ACT Prep Guide 2016–2017* that demonstrate the key concepts. All problems are listed with a three-number code, defined as

[test number].[section number].[question number]

Thus "1.2.15" is test 1, section 2, question 15 in the book. Section 2 is the math section, containing 60 questions. Note also that math questions are numbered by difficulty, so questions toward the end of a section are generally harder than those at the beginning.

Alignment Beyond the Test

The amount of alignment relevant to you depends on your goal for your class. If you are simply teaching test preparation for the sake of doing well on the tests, you should be concerned with the skills that are required to answer the questions and the skills that will be needed during the test prep classroom activities. These skills alone account for the majority of the Standards. This is an important point; you can be confident that *test prep is fully compatible with the Common Core* and helps students acquire and refine real skills they will need and use in college or careers.

Remember also that it's okay for some standards to align poorly. No one is claiming that these tests will perfectly satisfy all the Common Core requirements. Test preparation should not replace your usual math classes, but it can *supplement* them. Test prep gives you more drills, more data, and more resources for instruction, all of it fully compatible with the Standards. Not every Standard aligns with the tests, but all of the tests align with the Standards.

Missing Skills

Since the ACT is a high school test, we're focusing on the high school standards in this discussion. But there are also some concepts on the test that students should have learned *before* high school. For example, computing a simple percentage is a skill that shows up on the ACT in all sorts of questions, but percentages aren't mentioned in any of the standards here. However, computing percentages does show up in standards for sixth and seventh grade. So if you encounter an ACT question that doesn't seem to have a matching standard, it may align to an earlier standard.

Science Test Alignment

There is no specific set of Common Core Science Standards that aligns neatly to the ACT Science Test. The CCSS does not cover any specific science content, nor are science concepts referred to in any of the mathematics domains.

Luckily, the ACT Science Test also doesn't require any specific science content. It's much more about the skills of scientific thought, such as interpreting figures and making conclusions based on evidence. *Those* skills show up all over the Common Core. But they actually align more closely to standards found in the Common Core ELA standards. This may seem counterintuitive, but it makes sense given the nature of the test. The Science Test gives passages that students must interpret, not unlike what they must do on the Reading Test.

We won't go through all the ELA standards here, but at the end of this section, we will highlight the particular ELA standards that align to the Science Test.

Standards for Mathematical Practice [MP]

1. **Make sense of problems and persevere in solving them.**
2. **Reason abstractly and quantitatively.**
3. **Construct viable arguments and critique the reasoning of others.**
4. **Model with mathematics.**
5. **Use appropriate tools strategically.**
6. **Attend to precision.**
7. **Look for and make use of structure.**
8. **Look for and express regularity in repeated reasoning.**

Alignment

The ACT is fully aligned with all Standards for Mathematical Practice, except standard MP.5, to which it is not aligned.

Discussion

The Standards for Mathematical Practice are well aligned to the ACT. These skills are necessary for questions throughout the test material. In fact, it's skills such as these that can make the test so difficult for so many students, and they were at the forefront of our thoughts when we designed our test preparation system.

Many of these standards are directly or indirectly addressed in the "Math Techniques" chapter of the *Book of Knowledge*. The "General Strategies" section outlines strategies to help students piece together confusing problems and understand confusing problems. It has strategies to help students read carefully and eliminate careless mistakes. It has strategies that help students switch between abstract reasoning and quantitative reasoning. It has strategies with concrete structures that are repeatable across many different question types. In short, it specifically addresses almost all of the standards listed here.

Let's go through all of them one by one.

1. Make sense of problems and persevere in solving them.
2. Reason abstractly and quantitatively.

These are obviously true on their face for all ACT Math problems. This is what it means to solve math problems.

3. Construct viable arguments and critique the reasoning of others.

The ACT considered this to be aligned, but it's a bit of a stretch. The second part, about critiquing others, doesn't really apply to ACT Math problems. Science problems, sure. Reading problems, ok. Math? Not so much. The first part, "construct viable arguments", is applicable, albeit somewhat. Students won't be asked to literally construct arguments. There are no proofs, and you don't have to justify your answer. However, students will have to reason and form conclusions, reason inductively about data, and determine multistep procedures for solving a problem.

4. Model with mathematics.

There is a significant amount of mathematical modeling on the ACT, including but not limited to creating functions based on contexts in plain English and making relationships between functions, tables, and graphs.

5. Use appropriate tools strategically.

The ACT document states that the test does not align with standard MP.5, along with the note, "The College Readiness Standards emphasize many of the decisions students must make about approaching problems, although technology is not covered explicitly." This may seem odd, since the ACT permits the use of calculators (even some advanced graphing calculators), which would certainly qualify as a "tool". But the ACT is *calculator-optional*: Calculators are permitted but *never required*. The test does not care what method students use to solve a problem, nor does it care whether or not they use tools. As such, the document considered the test to be unaligned to MP.5.

That said, even if this standard isn't aligned by necessity, it is aligned in practice and should be a part of any prep work. Students are permitted calculators, and there's really no reason why they shouldn't have them for the test. Learning when it is and is not advantageous to use a calculator is an important part of the test.

6. Attend to precision.

Yes, yes, yes, absolutely. Carelessness is a huge problem for some students. They have a solid foundation of mathematical knowledge but constantly make small mistakes and get questions wrong, often because of timing concerns. Questions are written with distractor choices that are intentionally tempting and trip up students who are not careful.

7. Look for and make use of structure.

ACT questions often test structure, explicitly and implicitly. Algebra questions in particular will ask students to rewrite the structure of equations in order to highlight a particular quantity or a feature of a graph. Indirectly, more complicated questions will require students to see structural similarities or connections that aren't immediately apparent, such as treating multiple variables together as single units, or rearranging given equations to find familiar identities.

8. Look for and express regularity in repeated reasoning.

ACT problems often use the same kinds of structures in multiple different problems. While questions do not require particular methods, students should frequently be able to apply the same patterns to multiple problems, and being able to recognize that repetition is an important skill on the test. Furthermore, ACT problems often explicitly ask students to notice patterns. For example, questions may present a table of values and ask students to determine if they represent linear or exponential growth, or to write an equation describing the relationship between the variables.

Summary

The ACT is strongly aligned with the Standards for Mathematical Practice. Standard MP.5 is not strictly aligned because calculators are optional, but effective use of calculators is an important part of the test. Furthermore, A-List's math techniques encourage the development and refinement of these very skills.

Number and Quantity [N]

The Real Number System [N-RN]

- *Extend the properties of exponents to rational exponents. (2)*
- *Use properties of rational and irrational numbers. (1)*

Quantities [N-Q]

- *Reason quantitatively and use units to solve problems. (3)*

The Complex Number System [N-CN]

- *Perform arithmetic operations with complex numbers. (3)*
- *Represent complex numbers and their operations on the complex plane. (3)*
- *Use complex numbers in polynomial identities and equations. (3)*

Vector and Matrix Quantities [N-VM]

- *Represent and model with vector quantities. (3)*
- *Perform operations on vectors. (2, 5 subskills)*
- *Perform operations on matrices and use matrices in applications. (7)*

Alignment

The ACT is fully aligned with all standards in the Number and Quantity conceptual category.

Discussion

What Aligns

The ACT contains a number of questions that explicitly deal with the real number system and quantities. The standards within these two domains align with the ACT College Readiness Standards of Number and Quantity and Statistics and Probability. The ACT will contain problems involving properties of exponents, both integer and rational exponents, and will require students to know and employ rules for their manipulation.

The test will contain problems that require quantitative reasoning and use of appropriate units. Certainly, it is extremely common for problems on the test to employ units, but students will also encounter problems in which they have to convert units (often with conversion ratios provided) or interpret graphs that contain disparate or nonlinear scales on their axes.

Irrational numbers are rarely the central concept of a problem—questions don't tend to explicitly test understanding the idea of irrationality. However, the ACT will frequently make use of irrational numbers such as certain square roots or π.

The ACT may include some questions involving complex numbers (including manipulation of complex polynomials and graphing on the complex plane) and basic operations of matrices. However, it should be noted that these questions are very infrequent—no more than one or two questions per test, if any at all.

What Doesn't Align

The ACT will not contain questions that explicitly deal with vectors in vector format, so it's somewhat surprising that the alignment document considers all of N-VM to be aligned. However, the test does include questions involving concepts that can be represented as vectors, such as velocity. This may be why the ACT considers the vector-based standards aligned, even though vectors have not explicitly appeared on the test. Furthermore, the skills needed to understand some of the concepts in these standards (such as representing vectors graphically) are similar to skills that do appear on the ACT in other context (such as graphing on the *xy*-coordinate plane). That is, if you're looking for serious vector work, you won't find it on the ACT, but you will see some of basic concepts and some overlapping skills. That said, recent tests have indicated that explicit vectors may become more common, as their new book of practice contains several vector questions. So this may be changing in the coming years.

Summary

Alignment

The ACT will contain problems involving exponents with both integer and rational values and will make use of rational and irrational numbers. The ACT often requires quantitative reasoning. The ACT will contain problems involving complex numbers or matrices, albeit in small numbers. The ACT will generally not test knowledge of vectors, but it will use concepts that can be represented as vectors.

Sample ACT Questions

Exponents: 1.2.53, 1.2.56, 2.2.36, 3.2.48

Irrational Number Properties: 1.2.6

Quantities: 1.2.55, 2.2.45

Complex Numbers: 1.2.42, 2.2.56, 3.2.41

Matrices: 1.2.48, 2.2.42

Vectors: 2.2.57, 3.2.54

Algebra [A]

Seeing Structure in Expressions [A-SSE]

- *Interpret the structure of expressions. (2, 2 subskills)*
- *Write expressions in equivalent forms to solve problems. (2, 3 subskills)*

Arithmetic with Polynomials and Rational Expressions [A-APR]

- *Perform arithmetic operations on polynomials. (1)*
- *Understand the relationship between zeros and factors of polynomials. (2)*
- *Use polynomial identities to solve problems. (2)*
- *Rewrite rational expressions. (2)*

Creating Equations [A-CED]

- *Create equations that describe numbers or relationships. (4)*

Reasoning with Equations and Inequalities [A-REI]

- *Understand solving equations as a process of reasoning and explain the reasoning. (2)*
- *Solve equations and inequalities in one variable. (2, 2 subskills)*
- *Solve systems of equations. (5)*
- *Represent and solve equations and inequalities graphically. (3)*

Alignment

The ACT is aligned with all standards in the Algebra conceptual category, with partial alignment to two standards, A-REI.9 and A-REI.11.

Discussion

What Aligns

The ACT includes a huge variety of algebra questions in substantial quantities: anywhere from a third to half of the questions will deal directly with algebraic concepts. These concepts include:

- solving simple one-variable equations,
- manipulating multivariable equations and expressions,
- solving multivariable systems of equations,
- manipulating and understanding polynomials,
- simplifying rational expressions,
- solving inequalities, and
- creating and solving equations based on settings expressed in words.

That last concept is particularly common on the test, and directly aligns with the kind of "modeling" that is so important to the Common Core Math Standards. All of these topics are discussed in the respective algebra sections of the *Book of Knowledge*. The ACT splits its algebra questions into two main categories, *Elementary Algebra* and *Intermediate Algebra*, which together make up 30–40% of the test. Furthermore, some of the concepts in these standards also may appear in the ACT's *Pre-Algebra* and *Coordinate Geometry* categories, which make up another 35–45% of the test.

Additionally, the "Math Techniques" chapters in the *Book of Knowledge* give methods of solving problems specifically geared toward students who struggle with algebra questions, either through not understanding the concepts, working too slowly, or making careless mistakes. All of our techniques provide a way for students to develop an understanding of how the abstraction of algebra is a generalization of concrete solutions.

What Doesn't Align

The ACT is partially aligned with standards A-REI.9 and A-REI.11. In both cases, the unaligned part of the standard specified the method a student should use to solve a problem (i.e., "using technology"). As mentioned before, while the ACT permits using technology in the form of calculators, the test does not require any particular method of solving a question.

Summary

The ACT is very strongly aligned to all the Algebra domains. The test does not demand that students use particular methods for solving problems, but nearly all the concepts listed in these standards are available to them.

Sample ACT Questions

Structure of Expressions: 1.2.15, 2.2.9, 3.2.21

Polynomials and Rational Expressions: 1.2.47, 2.2.28, 3.2.23

Creating Equations: 1.2.14, 2.2.5, 3.2.19

Reasoning with Equations and Inequalities: 1.2.49, 2.2.12, 3.2.32

Functions [F]

Interpreting Functions [F-IF]

- *Understand the concept of a function and use function notation. (3)*
- *Interpret functions that arise in applications in terms of the context. (3)*
- *Analyze functions using different representations. (3, 7 subskills)*

Building Functions [F-BF]

- *Build a function that models a relationship between two quantities. (2, 3 subskills)*
- *Build new functions from existing functions. (3, 4 subskills)*

Linear, Quadratic, and Exponential Models [F-LE]

- *Construct and compare linear, quadratic, and exponential models and solve problems. (4, 3 subskills)*
- *Interpret expressions for functions in terms of the situation they model. (1)*

Trigonometric Functions [F-TF]

- *Extend the domain of trigonometric functions using the unit circle. (4)*
- *Model periodic phenomena with trigonometric functions. (3)*
- *Prove and apply trigonometric identities. (2)*

Alignment

The ACT is aligned with all standards in all four domains of the Functions conceptual category, with partial alignment to three standards: F-IF.7, F-LE.4, and F-TF.7.

Discussion

What Aligns

The ACT has a substantial number of questions that explicitly test knowledge of functions and function notation. These questions come in several different forms. Some will introduce an unknown symbol and provide a set of procedures that defines that symbol. Some will provide equations in *f*(*x*) format, asking students to evaluate, manipulate, or transform those equations. Some will provide similar equations in the context of an actual setting. Others still will ask students to produce, interpret, or transform graphs of functions. Students will certainly be required to understand how functions work.

The ACT is fully aligned with all the Trigonometric Functions standards. As a rule, each test will contain three to six questions specifically about trigonometry and associated functions. These questions can range in difficulty, including simple computation of trigonometric functions with respect to a triangle, questions that require students to construct functions based on specific settings, and graphs of major functions and the unit circle. Similarly, the ACT will have some questions that deal with logarithms (though not in such strictly defined quantities).

Furthermore, it should be noted that A-List's math techniques like Plug In and Backsolve work just as well for function problems as they do for algebra problems. These techniques can be particularly useful for intricate questions for which students easily get lost, such as those involving inverse functions, nested functions, or function transformations.

What Doesn't Align

The ACT alignment document states that the test has a partial alignment with F-IF.7, F-LE.4, and F-TF.7 because of mentions of "technology" in each. This is the standard warning that the ACT gives for all standards that mention technology, since the test does not care what method students use to solve the questions, only whether the students produce the correct answers.

Summary

The Function category aligns strongly with the ACT. The test frequently contains questions that test knowledge of functions in many different ways.

Sample ACT Questions

Interpreting Functions: 1.2.18, 2.2.54, 3.2.11

Building Functions: 1.2.20, 2.2.23, 3.2.49

Linear, Quadratic, and Exponential Models: 1.2.44, 2.2.34, 3.2.45

Trigonometric Functions: 1.2.39, 2.2.48, 3.2.23

Geometry [G]

Congruence [G-CO]

- *Experiment with transformations in the plane. (5)*
- *Understand congruence in terms of rigid motions. (3)*
- *Prove geometric theorems. (3)*
- *Make geometric constructions. (2)*

Similarity, Right Triangles, and Trigonometry [G-SRT]

- *Understand similarity in terms of similarity transformations. (3, 2 subskills)*
- *Prove theorems involving similarity. (2)*
- *Define trigonometric ratios and solve problems involving right triangles. (3)*
- *Apply trigonometry to general triangles. (3)*

Circles [G-C]

- *Understand and apply theorems about circles. (4)*
- *Find arc lengths and areas of sectors of circles. (1)*

Expressing Geometric Properties with Equations [G-GPE]

- *Translate between the geometric description and the equation for a conic section. (3)*
- *Use coordinates to prove simple geometric theorems algebraically. (4)*

Geometric Measurement and Dimension [G-GMD]

- *Explain volume formulas and use them to solve problems. (3)*
- *Visualize relationships between two-dimensional and three-dimensional objects. (1)*

Modeling with Geometry [G-MG]

- *Apply geometric concepts in modeling situations. (3)*

Alignment

The ACT is aligned with all standards in the Geometry conceptual category, with partial alignment to three standards, G-CO.2, G-CO.5, and G.CO.12

Discussion

What Aligns

The ACT has a significant number of geometry questions, with 35–45% of the test consisting of Plane and Coordinate Geometry questions. The standards as they're shown here are a fairly effective summary of the major topics covered on the tests: Angles, Triangles, Circles, and Volume.

The prevalence of geometry is one of the marked differences between the ACT and the SAT. The old SAT used to test geometry at frequency similar to the ACT, but the redesigned test has cut geometry back to less than 10% of the test. In contrast, Plane Geometry is tied with Pre-Algebra for the most prevalent concept on the ACT. If you want geometry, the ACT is the test for you. Unlike the SAT, the ACT does not provide any geometric formulas to students. Whereas students on the SAT may look up select formulas for area and volume, ACT-takers must have these formulas memorized.

The ACT will test trigonometry, both in the context of trigonometric functions (as described in the Functions section above) and in the context of formulas such as the law of sines or cosines. In fact, when such laws are necessary for an ACT problem, the question often gives the students the formula for the law in the text of the question itself (unlike area and volume formulas).

Geometry questions also have a special place for one of A-List's math techniques. Our two main techniques, Plug In and Backsolve, are generally used in an algebraic context, though they often work just as well for geometry problems. But our third technique, Guesstimate, is specifically designed for geometry. It asks students to use their skill at observing and visually evaluating measurements directly in order to approximate the answer in increments of increasing accuracy. That is, look at the picture and guess the size of the quantity you want, relative to the quantities you know. While the ACT does not guarantee that all figures are drawn to scale, we have found that in practice, figures are drawn to scale 95% of the time, so Guesstimate is very much an option.

What Doesn't Align

The ACT's partial alignment, as always, is the result of standards that specify the use of software or other technology. As mentioned, the ACT allows calculators but is ultimately indifferent to the methods students use.

The geometry standards frequently ask students to construct proofs of certain theorems or to derive particular formulas using other theorems or rules. This kind of work is grounded in the classical Euclidean tradition of geometry and is often the focus of geometry classes in school. The ACT, however, is a results-based test and is only concerned with the solution to a particular problem. Thus students will never be asked to explicitly prove or derive any geometrical proof. That's not to say that these skills are unimportant for the test. On the contrary, this test, rather than asking students to know a lot of things superficially, asks students to know a *few* things *really well*. Indeed, this is why many students struggle with the test. Therefore, being able to prove theorems rather than simply memorize them can give students a much deeper understanding of the underlying truths, and will make employing the theorems when necessary that much easier.

Summary

The ACT contains a significant number of geometry questions in all the domains listed above. The ACT will include trigonometry but will not require students to prove or derive any formulas or theorems, but the ability to do so can give students a deeper understanding of the necessary geometric principles.

Sample ACT Questions

Congruence: 1.2.16, 2.2.50, 3.2.44

Similarity, Right Triangles and Trigonometry: 1.2.30, 2.2.7, 3.2.20

Circles: 1.2.35, 2.2.10, 3.2.29

Expressing Geometric Properties with Equations: 2.2.38, 3.2.56

Geometric Measurement and Dimension: 2.2.13, 3.2.42, 3.2.43

Modeling with Geometry: 1.2.17, 2.2.21, 3.2.25

Statistics and Probability [S]

Interpreting Categorical and Quantitative Data [S-ID]

- *Summarize, represent, and interpret data on a single count or measurement variable. (4)*
- *Summarize, represent, and interpret data on two categorical and quantitative variables. (2, 3 subskills)*
- *Interpret linear models. (3)*

Making Inferences and Justifying Conclusions [S-IC]

- *Understand and evaluate random processes underlying statistical experiments. (2)*
- *Make inferences and justify conclusions from sample surveys, experiments, and observational studies. (4)*

Conditional Probability and the Rules of Probability [S-CP]

- *Understand independence and conditional probability and use them to interpret data. (5)*
- *Use the rules of probability to compute probabilities of compound events in a uniform probability model. (4)*

Using Probability to Make Decisions [S-MD]

- *Calculate expected values and use them to solve problems. (4)*
- *Use probability to evaluate outcomes of decisions. (3, 2 subskills)*

Alignment

The ACT is aligned with all standards in all four domains of the Statistics and Probability conceptual category.

Discussion

What Aligns

The ACT contains a number of questions dealing with concepts of Statistics and Probability. The ACT does not have a separate category of questions dedicated to statistics, but most of the concepts listed in these standards—such as median, mean, and probability—fall into the "Pre-Algebra" category, which collectively constitutes 20–25% of the test.

Median, mean, and probability are frequently present on the ACT. The test will also contain questions featuring various ways of presenting data. These could be any of a wide variety of formats, including simple tables, bar graphs, scatterplots, pictographs, stem and leaf plots, and Venn diagrams.

What Doesn't Align

The ACT's alignment document can be a bit liberal about whether a standard should be considered aligned. For example, in the Algebra standards, the ACT considers itself to be aligned with A-APR.5: "Know and apply the Binomial Theorem for the expansion of $(x + y)^n$ in powers of x and y for a positive integer n, where x and y are any numbers, with coefficients determined for example by Pascal's Triangle." We at A-List have yet to see any question on the ACT that dealt with Pascal's Triangle explicitly, but there are questions involving binomial expansion with a value of $n = 2$. Technically, those $n = 2$ questions could be considered aligned with this standard, even if they may not go quite as far as the standard likely intends.

This problem is most noticeable in the Statistics and Probability standards. Historically, statistics and probability have not been as prevalent on the ACT the alignment documents would lead you to believe. Many of the standards here seem to ask for a far greater level of data interpretation than the test's math section demands of its students. The majority of such questions on the tests involve little more than *reading* a graph, or *computing* a probability, with very little actual *analysis* of the data demanded by, say, standard S-ID.9, "Distinguish between correlation and causation."

However, this seems to be changing. Their most recent book has shown a marked increase in the number of explicit statistics questions, including scatterplots and expected values based on probabilities. Starting in the fall of 2016, statistics will now be a reporting category on students' score reports. This was not the case when the alignment document was written, but the test seems to be making a conscious effort to include more statistics.

Additionally, ACT has a number of questions involving more in-depth data analysis in *the Science Test*. This test is composed of six to seven passages, each of which displays a number of figures, tables, and graphs surrounding an experiment or scientific situation. The questions then ask students to be able to read data from the figures, extrapolate and predict data points not present, describe relationships between variables, and draw conclusions based on the data. These questions get closer to the level of statistical manipulation that the standards in these domains demand. If you have particular interest in focusing on statistics with your students, don't be limited by the Math Test. The Science Test may be better suited for your purposes.

Summary

The ACT aligns with the Statistics and Probability category and contains a large number of problems dealing with median and mean, data interpretation, and probability. The test previously did not include statistics in significant quantities, but its presence will increase going forward. The ACT Science section provides an additional source of questions about data interpretation.

Sample ACT Questions

Mean, Median, and Mode: 1.2.41, 2.2.29, 3.2.35

Probability: 1.2.54, 2.2.47, 3.2.46

Modeling from Data: 1.2.44, 2.2.25, 3.2.59

ACT Science Alignment

As we mentioned earlier, there is no set of Common Core Science Standards that can be easily matched up with the ACT Science Test. There's very little actual math on the Science Test, so it generally does not align with any of the math standards. We did see some mention of the Science Test in reference to certain Statistics and Probability standards, but that's all. However, that doesn't mean that the test doesn't align to the CCSS. Rather, the test aligns with some of the ELA standards, particularly in the Reading strand.

Since ELA is not the focus of this book, we won't go through the entire set of ELA standards here. Rather, we'll just highlight those standards that align well to ACT Science. Note also that some of the standards listed here may also align with the Reading Test, but we will not go into discussions of Reading here. For more on Reading and alignment to the ELA standards, see our book, *ACT ELA in the Classroom.*

How to Read ELA Standards

The ELA standards are organized slightly differently from the math standards, so let's take a minute to explain what's going on here.

The ELA standards are divided into strands (Reading, Writing, Language, etc.). Each strand has a list of College and Career Readiness (CCR) Anchor Standards. These are the broad standards that students must attain in order to be deemed fit for college or a career beyond high school.

In addition, these standards are broken out for each grade level from kindergarten to grade 12, outlining what a student must be able to do and understand by the end of that grade level. For high school, standards are shown for two groupings, grades 9–10 and grades 11–12, but the alignment document only focused on the 11–12 standards.

For Reading, these grade-level standards are broken out into different content areas. That is, Reading has one set of Anchor Standards but has different grade-level standards for Reading Literature (RL), Reading Informational Texts (RI), Reading for Literacy in History/Social Studies (RH), and Reading for Literacy in Science and Technical Subjects (RST). Most of these do not align well with the Science Test, with the exception of RST, which will be discussed in its own section.

Each standard has a code containing three parts: letters denoting the strand or content area, the grade level of the standard or "CCR" for Anchor Standards, and the sequential number of the standard (sometimes with letters for subpoints). For example, "RL.CCR.5" refers to the fifth Anchor Standard in the Reading Literature strand.

In the tables presented here, the last column shows a Y if the standard is aligned, an N if it is not aligned, and a P if it is partially aligned. An S means the standard is aligned to the Science Test, while an R means it is aligned to the Reading Test. Otherwise, the overall layout of this section is the same as the previous sections.

What Aligns with Science

The following standards align to the Science Test:

- Reading Anchor Standards
  - R.CCR.7
  - R.CCR.8
  - R.CCR.9
- Reading Informational Text
  - RI.11–12.7
- Reading for Literacy in Science and Technical Subjects
  - RST.11–12.3
  - RST.11–12.7
  - RST.11–12.8
  - RST.11–12.9
  - RST.11–12.10

Reading Anchor Standards and Informational Text

These three Anchor Standards are all grouped together under the heading "Integration of Knowledge and Ideas". Each of the Anchor Standards has accompanying grade-level standards in the Reading Informational Text strand. Most of those do not align because they make specific references to historical texts that are outside the scope of the Science Test. But we do include Informational Text standard 7, which does align.

Code	Standard	Aligns
R.CCR.7	**Integrate and evaluate content presented in diverse formats and media, including visually and quantitatively, as well as in words.**	**P**
RI.11–12.7	Integrate and evaluate multiple sources of information presented in different media or formats (e.g., visually, quantitatively) as well as in words in order to address a question or solve a problem.	**P**
R.CCR.8	**Delineate and evaluate the argument and specific claims in a text, including the validity of the reasoning as well as the relevance and sufficiency of the evidence.**	**Y**
R.CCR.9	**Analyze how two or more texts address similar themes or topics in order to build knowledge or to compare the approaches the authors take.**	**Y**

Alignment

The CCR Anchor Standards 7, 8, and 9, and Informational Text standard 7 align with the ACT Science Test. R.CCR.7 and RI.11–12.7 are partially aligned, excluding the words "and media".

Discussion

Standard 7

This standard precisely describes the ACT Science Test, which is entirely about quantitatively evaluating content. Despite the name, it does not primarily test students on their knowledge of scientific facts. Rather, it is composed of passages that contain some combination of tables, graphs, figures, and explanatory texts. These figures present data in a wide variety of different formats: line graphs, shaded area graphs, scatterplots, diagrams, and all sorts of unusual combinations of axes.

The grand majority of the questions involve evaluating visual or quantitative format in the context of a given topic. Some questions ask students to directly extract data from the tables and graphs. Others ask them to compare and combine data from multiple figures, or to combine data from the figures with information in the explanatory text.

In this way, the ACT Science Test is very similar to the ACT Reading text. It does contain a great deal of science content, but everything the student needs to know and understand is defined and explained in the passage itself. The questions focus on the students' ability to find and evaluate information given in the text as well as in a variety of visual and quantitative figures.

The Informational Text standard here is very similar to the Anchor Standard, so it aligns just as well.

Standard 8

The ACT Science Test often has what we call *Reasoning Questions*. These are questions that present a certain thesis and ask whether the information in the tables and figures support that conclusion. These questions demand not only a yes or no answer, but ask students to supply appropriate evidence to support that conclusion. When doing these questions, then, students must answer two parts. First, whether the evidence in a choice is actually true according to the data provided in the passage. Second, if it is true they must determine whether that evidence supports or weakens the stated conclusion.

Additionally, Conflicting Viewpoints passages present a problem or situation followed by two or more hypotheses that attempt to explain it. The questions often focus on the students' ability to follow the logic of an argument and contrast opposing points of view. For example, a question may ask "which of the following, if true, would weaken Hypothesis 1?" This does not require the student to actually take a position, but it does require the student to determine what facts would or would not affect the logical force of the argument. Or a question might present a new fact and ask "which, if any, of the hypotheses is strengthened by this fact?"

Standard 9

As we mentioned ACT Science Test will contain one Conflicting Viewpoint passage. These passages present multiple viewpoints describing the same issue or topic. Questions will often ask about the relationship between the viewpoints, where they agree and disagree, which data they make use of, which data strengthens or weakens their positions and the relative strengths of their arguments.

What Doesn't Align

For standard 7, the tests do not align with presentations in other "media". The ACT will only be offered on paper. There will be no digital, audio, or video components to the test.

For standard 9, the questions will be concerned only with the content of the arguments, not the nature or quality of the author's writing. The questions often delve more deeply into the logic of the hypotheses argument.

Summary

The ACT includes questions with varied visual and quantitative formats on the Science Test.

The ACT Science may ask questions about conclusions that can be drawn from data, and in particular will evaluate the strength of arguments made in Conflicting Viewpoints passages.

The ACT Science Test contains one Conflicting Viewpoints passage in which students must compare multiple hypotheses on a single topic.

Sample ACT Questions

Figure Types: Any ACT Science Test will show a wide array of different figures and tables. For example, compare the following figures from test 1 in *The Official ACT Prep Book*:

Passage I, Figure 1: An anatomical diagram of a deer mouse's fur pigmentation along with four line graphs.

Passage II, Figure 1: A drawing illustrating the set up of an experiment described.

Passage IV, Figure 1: a bar graph with two axes.

Passage V, Table 1: a table with six rows and four columns, including numbers and arrows showing directions.

Reasoning Questions: 1.4.11, 2.4.12, 3.4.4

Conflicting Viewpoint Passages: 1.4.14–20, 2.4.28–34, 3.4.27–33

Reading for Literacy in Science and Technical Subjects

Below are listed all the standards in the Reading for Literacy in Science and Technical Subjects grade-level strand. Per the alignment document, any standard here that doesn't align to the Science Test does align to the Reading Test, particularly to the natural science passage.

Code	Standard	Aligns
Key Ideas and Details		
RST.11–12.1	Cite specific textual evidence to support analysis of science and technical texts, attending to important distinctions the author makes and to any gaps or inconsistencies in the account.	PR
RST.11–12.2	Determine the central ideas or conclusions of a text; summarize complex concepts, processes or information presented in a text by paraphrasing them in simpler but still accurate terms.	YR
RST.11–12.3	Follow precisely a complex multistep procedure when carrying out experiments, taking measurements, or performing technical tasks; analyze the specific results based on explanations in the text.	**YS**
Craft and Structure		
RST.11–12.4	Determine the meaning of symbols, key terms, and other domain-specific words and phrases as they are used in a specific scientific or technical context relevant to grades 11–12 texts and topics.	PR
RST.11–12.5	Analyze how the text structures information or ideas into categories or hierarchies, demonstrating understanding of the information or ideas.	YR
RST.11–12.6	Analyze the author's purpose in providing an explanation, describing a procedure, or discussing an experiment in a text, identifying important issues that remain unresolved.	YR
Integration of Knowledge and Ideas		
RST.11–12.7	Integrate and evaluate multiple sources of information presented in diverse formats and media (e.g., quantitative data, video, multimedia) in order to address a question or solve a problem.	**PS**
RST.11–12.8	Evaluate the hypotheses, data, analysis, and conclusions in a science or technical text, verifying the data when possible and corroborating or challenging conclusions with other sources of information.	**YS**

(Continued)

Code	Standard	Aligns
RST.11–12.9	Synthesize information from a range of sources (e.g. texts, experiments, simulations) into a coherent understanding of a process, phenomenon, or concept, resolving conflicting information when possible.	**YS**
Range of Reading and Level of Text Complexity		
RST.11–12.10	By the end of grade 12, read and comprehend science/technical texts in the grades 11-CCR text complexity band independently and proficiently.	**YRS**

Alignment

The ACT Science Test is aligned with standards 3, 7, 8, 9, and 10, with the exception of the phrases "and media" and "video, multimedia" in standard 7.

Discussion

What Aligns

These standards cover a lot of the same ground we saw in the Anchor Standards but now specifically placed into a scientific context. The added context means that some concepts that weren't aligned for the Anchor Standards are now aligned when applied to science.

The ACT Science Test contains passages that present original science material. Each passage contains a textual introduction which is accompanied by a variety of tables and graphs—information presented in multiple formats (standard 7)—that the students must analyze and evaluate directly on their own merits (standard 8). Research Summaries passages present detailed multistep procedures (standard 3), and students must answer questions about their design and results. Furthermore, the Conflicting Viewpoints passage is specifically designed to present two or more differing hypotheses on a given topic or situation, followed by questions asking students to compare the arguments and attempt to resolve discrepancies (standards 7, 8, and 9).

What Doesn't Align

As we saw in the Anchor Standards, standard 7 partially aligns, excluding the phrases "and media" and "video, multimedia". The ACT is given only on paper and will not contain video or other media.

According to the alignment document, standards 1, 2, 4, 5, and 6 align to the Reading Test but not the Science Test. The main difference is that the Science Test is more likely to ask about the scientific content of a passage, while the Reading Test is more likely to ask about the structure and prose of the passage. But this seems like an unfair exclusion; these reading standards do seem relevant to the Science Test as well. Standard 2, for example, seems like the kind of thing that is asked of students on certain types of Conflicting Viewpoints passages, when asked to understand the respective hypotheses presented. Standard 4 involves determining and understanding key technical terms. While questions rarely ask students to explicitly define terms, students will have to understand unfamiliar terms that are presented and defined in the passage, based on the context in which they are given. These standards perhaps do not align as strongly as those discussed above, but they are relevant to the skills used in the Science Test.

Summary

Between the Science Test and the natural science passage on the Reading Test, the ACT is aligned with all of these standards. The Reading Test will ask questions about how the central ideas and construction of the text itself. The Science Test will ask about the data, experiments, and conclusions contained in a text.

3

ACT Math in the Classroom

Overview

We know that ACT Math correlates with Common Core Math. What does that mean for you the teacher? How can you get your math classes ready for the ACT?

Some of you may have a full ACT prep course set up as its own class in your school, either as an elective or after school. If you have time to dedicate exclusively to test prep, it's best to get some materials specifically designed for such a course. A-List not only has our own materials for such programs, we also offer professional development and assistance in setting up your courses.

Full prep classes should be reserved for junior year. You want to make sure a) that they've been exposed to all the math they'll need for the test, and b) that the classes are given close to the time when students will take the real test. If a prep class ends in June but the students don't take the real test until October, they will surely forget everything you've done in the meantime.

Not everyone has the time and resources available for dedicated prep courses. But that's okay! This section will show you how you can weave ACT math material into your normal math classes in ways that can enhance the material you're already using.

When you're preparing students for the ACT, there are three areas of the test you should consider *content*, *technique*, and *timing*. Then, we'll go through a number of ways to expose your students to practice problems, depending on which of these three areas you want to focus on.

Content

The first thing you can do is *don't change a thing*. Unless you're teaching calculus or other concepts beyond algebra 2, everything you're already doing is going to be relevant for the ACT to some degree. Not all concepts will be equally present, but that's inevitable with the constraints of the test. It's tough to squeeze three years of math into 60 questions. But just about everything you do will contribute somehow.

The easiest thing to do outside of a prep course is to simply weave sample ACT material into your courses. You can do so with material reflecting the work you're currently doing anyway, with work you've done recently this year, or surprise them with work from previous years.

How much you do is up to you. If you're working on content that is underrepresented on the ACT, then you'll probably want to scale back the amount of material. Trigonometry, for example, is less than 10% of the test. When you're teaching a trigonometry course, the stuff you're working on will show up, just not very often. So an occasional problem or two is useful, but not every day. If you're doing algebra of linear functions, that stuff is all over the test.

Of course you don't have to give problems that are directly relevant to today's lesson. It's great to call back to math concepts they've already learned (and forgotten). Paradoxically, sometimes students struggle more with the *easier* math because it's been too long since they've had to think about it. You've just spent the last week talking about parabolas, so they're fresh in your students' minds, but it's been years since anyone has asked them to add fractions. As students approach the end of high school, they remember less and less of the beginning of high school. We've seen so many students miss questions that require little more than sixth-grade math.

The other side of this is that the test may include some content that isn't in your curriculum at all. This may be true, for example, if your school doesn't have a dedicated statistics course, or if some more advanced algebra or trigonometry concepts don't show up until senior year. If that's the case, you may consider bringing them up specifically in the context of the ACT. Statistics, for example, show up on the test in a variety of contexts, both on the Math and the Science Tests, but never in great detail. You may see a scatterplot with a line of best fit, but you'll simply have to know what that means. You won't have to draw the line or calculate R-squared.

Techniques

Content is only one part of the test. There's also a question of *techniques*, figuring out the best way of doing problems. A-List uses two main math techniques: Plug In and Backsolve. They both have the same goal: to turn algebra problems into arithmetic problems.

- Plug In exploits the fact that an equation is true for all inputs (usually). If the answer choices contain algebraic expressions, then the right answer will always be right, no matter what specific values you choose. So pick a number for your variable, put it through the problem, and get another number as your answer. Put your made up number in the choices and see which gives you the same output.
- Backsolve works for questions where there only is one possible value. If there are numbers in the answer choices, pick a choice and make that number the answer to the question. Put the number through all the information you have and make sure everything matches. If it does, that's your answer. If it doesn't, pick a different choice.

Our textbook, *The Book of Knowledge*, discusses the math techniques in greater detail. At heart they are rather straightforward and don't take long to understand. But they do require a lot of practice to do well. This is true of literally all new skills. Just because you can read sheet music doesn't mean you can play the piano. You have to practice in order to execute.

You can introduce these techniques to your students in just a few minutes of class time. Once you do, you can bring them back up constantly. Any time you do a practice problem, ask if it can be done with a technique. Sometimes the answer is no, but overall about 40% of ACT questions can be done with Plug In or Backsolve. It takes practice to get comfortable using methods beyond what you're used to, but these methods are crucial to score improvement. If your students don't change their methods, they won't change their scores.

Timing

The last element of the test is putting the content and techniques together under a time limit. So many students have problems that stem from time management. The only way to work on these issues is with some timed practice.

Timed Practice

Full Tests

The best way to expose students to the test is to give real, proctored practice tests. Any exposure to practice problems is valuable for students, but full-length proctored tests are best for a number of reasons:

- They give students exposure to the test format and conditions.
- Practice tests given under real-test conditions produce scores that most accurately predict real scores.
- They allow students to practice timing and time-sensitive test-taking techniques.

Timing can be a huge issue for students, and even if they know all the math perfectly it can be difficult for them to acclimate to the high-pressure three-hour test format. Anyone with experience teaching the test has seen students who do much better on problems done leisurely without time constraints than on a timed section. The most common problems that arise include:

- Students move too slowly and run out of time before completing as many questions as they want.
- Students move too quickly and make careless mistakes on questions they know how to do.
- Pure mental fatigue of working for three hours negatively affects students' performance.

This is often what holds back students who normally do well in school but don't score as highly as they expect on the ACT. It's not that they don't know enough math; it's that they're missing questions they already know how to do. They know most of the math, but they struggle to apply it efficiently.

So what do we do about timing problems? It's something you can discuss explicitly with your students, particularly as you get closer to the real test. A-List's *Book of Knowledge* contains several discussions that are specifically geared toward time management. There are two main ways to address time management.

1. **Target numbers**. Unless you're trying to get a 36, you do not actually have to do every question. You'd be surprised how few questions you need in order to get the score you want. If you're trying to get, say, a 26 on the Math Test (about the 85th percentile), you only need to get about 43 questions right out of 60 available questions (that's 72%). If you do *fewer* questions but with *greater accuracy*, it's easier to get your target score. Because the questions are ordered by difficulty, you know you're more likely to get the questions at the beginning. Skip the last questions, spend more time on the first questions, and you'll cut down on careless mistakes. (Just remember to fill random answers for the questions you skip. You don't lose points for wrong answers so you'll pick up some points from random guessing. So, for example, if your goal is to get 43 questions, you could answer 40 of them and expect to get at least three of the last 20 right by random chance.)
2. **Math techniques**. As we mentioned earlier, the math techniques, Plug In and Backsolve, are ways to turn algebra problems into arithmetic problems. That often means making questions easier and making them go faster. More importantly, they help cut down on careless mistakes, so it makes you more accurate, and accuracy is our goal for using target numbers.

Math techniques can be practiced on any problem set, but in order to practice target numbers, you have to give timed sections.

PreACT

Starting in 2016, ACT will offer a new test called the PreACT to interested schools. This is effectively a preview of the ACT to be given to 10th grade students, much like the PSAT is a preview of the SAT. As of this writing, not very much specific information has been released about it, but ACT has said that it will contain the same kinds of problems and the same testing environment as the ACT. It will be graded on the same 1–36 point scale but with shorter testing times. Schools who sign up for it can offer it any time during the school year, from September to June.

Of course, the ACT used to have a different test called the PLAN that was intended as a preview of the ACT for 10th graders. It was discontinued after 2014, replaced with ACT Aspire. This is a longitudinal assessment system that can be used for students as early as third grade. It's an interesting system and is aligned to the same ACT College and Career Readiness Standards to which the ACT itself is aligned. However, these tests are very different from the ACT, with different question types and score scales, so they're not directly relevant as ACT practice *qua* practice. Apparently, more than a few people felt the same way, as the PreACT has quickly come to fill the hole left by the departed PLAN.

Regardless, the PreACT exists now, and it's a useful tool for schools that want to get early data on their students' readiness for the ACT. However, it is only available for 10th graders, so if you want to do prep work for 11th graders, it won't be an option for you.

Practice Sections in Class

Even if you are giving a full test-prep course in school, it can be tough to schedule full-length practice tests. As an alternative, you may consider having students take a full section in class. It should be easy to give a 35-minute science section within a single class period. However, a full math section takes 60 minutes, and you probably don't have that much time in one class.

You could break the section up over two days, 30 minutes each day. If you do that, it's probably a good idea to give them the *full* section for both days and let them split it up how they wish, rather than giving them separate packets of, say questions 1–30 one day and 31–60 the next. The reason is that the questions are ordered by difficulty—question 1 will be really easy and question 60 will be really hard. Therefore, students won't (and shouldn't) spend their time evenly throughout the section. They'll do the first ten questions a lot faster than the last ten questions. Instead, just be sure to give them 60 minutes total across the two days and allow them to divide their attention how they see fit.

When Should I Give Tests and Sections?

Practice tests can be given at any point in the year to any class of students, depending on the reason you're giving the test. You can give practice tests to students as young as ninth grade. As we've said, on a most basic level it gives students exposure to the format and a sense of where they are in relation to the national and state averages.

First off, you should be judicious about giving official practice tests to students. You don't want to give out all available practice ACTs to your sophomores so that they have none left to do as juniors.

For younger students, practice tests can be more instructive for administrative purposes than to help the students prep. Ninth graders won't do as well on practice tests as juniors will, and it may not be terribly useful to tell students that they aren't doing well on Algebra 2 concepts if they haven't taken Algebra 2 yet. But it can give the school valuable data about which areas the class as a whole is strong or weak in. This can be particularly important if your state uses the ACT as its statewide assessment and you want to see if your school is on track.

If you're giving a prep course, you should be sure to give a minimum of two practice tests: one at the beginning and one at the end. That way you can see how students have improved and you can gauge whether your course has been effective. The first test is just used as a baseline to see where students start. The second test should be toward the end of the course, close to the time when they'll take the real test. It should be given after you've gone over most of the necessary concepts. Of course, if students have already taken the real ACT once, you can use their prior scores as a baseline.

If you're not giving full proctored tests, only sections or partial tests given in class, you probably don't need to do a baseline test. Even if you do the whole test, if you do it in pieces, the scores won't be as reliable as they would be for a full-length timed test taken in one sitting. A baseline can still give some information, but that's a lot of class time to use just to generate scores that are unreliable.

For any test that isn't purely a baseline test, you should also spend some time going over the questions in class. Giving the questions is fine practice, but we also want students to improve their performance. To do that they need to see what they're doing wrong. It's not simply a matter of going over questions they missed; it's also a matter of making sure they're doing questions in the most effective way. In our experience, most students' score improvement comes not from lectures about math concepts and techniques, but from seeing those concepts and techniques in action when reviewing practice tests.

Since any section will contain questions from an assortment of different content areas, it's best to wait to give the practice until the students have been exposed to a majority of the concepts. It does little good to have them do problems that you know they'll struggle with at best, stare blankly into the void of space at worst. Once you've given them some grounding in the content, then you can start playing around with it in earnest.

Problem-a-Day

Of course, the simplest thing you can do is to give students a real ACT problem periodically in class. It doesn't have to take up a lot of time. It could just be a single question, once a week, a few times a week, or every day.

An ACT question is a great do-now activity, a nice, quick way to open a class. It doesn't take long; depending on the difficulty you can have the students do the question, review it, and give multiple ways of solving it in less than five minutes. Some teachers prefer to give a question at the end of class as an exit ticket. Either way, it needn't take much time out of your class.

This simple act accomplishes some important goals. By exposing students to the problems on a slow but regular basis, it familiarizes them with the test. On a base level, that makes it less scary for them, but it also habituates them to the sort of math they'll see. The test certainly repeats problem types over and over, so the more they see real problems, the more comfortable they'll be at solving them. It also reinforces the idea that ACT math *is not different* from regular math. It's not some special, elusive Fancy Math. It's just plain ol' math, like they do every day.

This sort of activity can be done with students of any age. It's a great thing to start on younger grades. If they see an ACT problem every day for three years, by the time they have to take the real test it will seem practically banal. Furthermore, you can demonstrate to them that the math they're doing in ninth grade is going to be important later on. If they can see that they'll need their ninth-grade math when they get to the ACT in eleventh grade, they're more likely to make and effort to remember it.

Which Problems Should I Choose?

Ideally, you can choose problems that use the same concepts you're talking about in class that day, or at least those you've been discussing recently. Been talking about graphing lines in the coordinate plate? The ACT *loves* graphing lines in the coordinate plane! Here, find this slope.

Of course, it's not always that easy. Sometimes it's tough to find a problem that exactly matches your current topic. Maybe you can find one, but not a whole week's worth. So if you want to commit to doing this on a regular basis, at some point, you'll likely have to go off-topic. That's okay. Just keep a few things in mind.

- As we've mentioned, don't do problems that are wildly above their skill level. If they're just starting Algebra 1, don't give them a problem where you have to divide polynomials and find the remainder.
- That said, it's sometimes nice to do problems involving new concepts if you can use that problem to introduce the concept. For example, say there's a problem dealing with an exponential function, which you haven't talked about yet. But the problem doesn't require complex computation or graphing, it just requires you to understand the meaning of the components of the equation. You can take this opportunity to show them the equation and explain what the parts mean. They likely won't get it right today, but many of them will the next time they see an exponential function, now that it's familiar to them.
- You may also decide to target your work based on data from test results. Whether that's from a practice test you administer yourself or from the real prior scores, any results will give you an abundance of information about your students' performance. Look for areas where they need the most help and come up with some targeted drills on those subjects.
- Be aware of the difficulty level of the problem. If it's a hard question you may want to give students more time than usual (and it may take you longer to go over it). It's okay to give a hard question to a remedial class; we firmly believe that anyone is capable of getting any question if given the right training and time constraints. But if it's a very hard question *and* it deals with a concept you haven't covered recently, you may want to save it until it's more directly relevant to your classwork. Note that while questions on practice tests are numbered by difficulty, questions acquired from other sources may not have difficulty information, so you'll have to use your judgment.

Where Can I Get Problems?

The best place to get problems is straight from ACT, Inc. There are several ways you can get full practice tests directly from the test-makers:

- A document called *Preparing for the ACT*, which includes one full-length practice test, can be downloaded for free from their website. Every three years, they make a new test available in this document. The current test as of 2015 is form 1572CPRE. Previous free tests, such as 1267C or 0964E, are no longer hosted on ACT's website, but your school may still have copies, and they can often still be found in other corners of the Internet.
- ACT, Inc. also publishes a book called *The Official ACT Prep Guide 2016–2017*, which contains three full-length practice tests.
- ACT also sells individual test booklets directly to schools and test prep organizations. There are two different forms available for sale to schools, currently forms 1467F and 1165D. Like the free download test, these forms are updated every three years, with the last update occurring in 2014.

But of course, you may not want to use the practice tests for daily classwork if you're saving them to be administered as proctored tests. Here are a few places where you can find additional practice material:

- *The Official ACT Prep Guide* comes with additional practice problems online. There are about 100 problems of each subject. It's just one big bank of questions, not organized as full test sections, but it's a good source to find practice.
- ACT offers an ACT Question of the Day service on their website. These are good questions, delivered daily. But it's pot luck what the topic is going to be, and it's not always a math question.
- A-List's ACT book, *The Book of Knowledge* is a thorough examination of the content of the new test and contains drills separated by content area along with lectures and

exercises to help do the problems effectively. A-List also has an online platform that offers a suite of supplemental material like quizzes and worksheets.

- ACT's website offers an *ACT Online Prep* course, which includes practice questions, games, and guided plans to help students prepare for the test. The course costs $39.95 for a one-year subscription. It has a large number of practice problems. However, the items are only available to view online or on mobile devices, not on paper. This can be problematic, since the real test is given on paper, and our techniques often rely on underlining and marking up passages. So the experience is not exactly the same.
- If you're really in a crunch, you could even use SATs. The format and timing are dramatically different, but most of the math problems are good math problems that are very similar to the ACT. The content isn't exactly the same, but it's very close; if you're looking for questions dealing with specific concepts, there's a good chance you can find some on the SAT. (The other major difference is that the SAT has four choices for the multiple-choice questions, not five like ACT Math.)

Translation Exercises

What if you want some middle ground? You're not giving a prep course and don't have time for full tests and sections, but you want to do more serious work in class than just a question a day. That's when you're going to want to have content-specific drills. You can find such drills in the *Book of Knowledge* or on *ACT Online Prep*. But you can also make your own drills using existing test material.

This is a process we call *translation*. Take an existing problem as a framework and rewrite it to make a new problem. This can be as simple as just changing the numbers or changing the setting. But you can also adjust it as you see fit. Add more concepts, trim it to fewer concepts, make it harder, make it easier, do what you want with it.

This is a valuable exercise not only because the output is a new problem set you can give your students but also because it forces you to analyze the problem deeply to understand the root concepts. It forces you to think about not just the correct answer but the incorrect answers. How will your students see this problem? Think about not just your solution, but your students' solutions.

There are several different ways you can go about doing this.

Direct Translation

You can translate questions to greater or lesser degrees, depending on your goal. Let's start with a simple problem:

(Note: The following problems are numbered sequentially for convenience. The numbers do *not* indicate difficulty level.)

1. **Bob has 4 dollars more than Lisa. If Lisa has x dollars, how much would Bob have if he doubled his money?**
   A. $x + 4$
   B. $x + 8$
   C. $2x$
   D. $2x + 4$
   E. $2x + 8$

Let's ignore the choices for now and focus on the question. The easiest way to change the question is just to change the names:

2. Sandeep has 4 dollars more than Carrie. If Carrie has *d* dollars, how much would Sandeep have if he doubled his money?

This is hardly any change at all. It's basically identical to the original problem. Let's not make it about money. Let's make it about something else.

3. Alan has 4 more cats than Lane has. If Lane has *c* cats, how much would Alan have if he doubled his cats?

Getting a little crazy, but okay. We're still talking about people owning objects, though. Let's try a different direction:

4. The Rockets won 4 more games than the Spurs. If the Spurs won *z* games, how many games would the Rockets win if they won twice as many games?

This starts out good, but ends up nonsensical. Moving into the realm of "wins" changes the problem conceptually to intangible objects. But as we can see, these intangible objects don't behave in quite the same way. Look at that: how many games would they win if they won twice as many games? What does that mean? We kept the underlying math the same, but we need to change the wording. We've got to be more explicit about the passage of time. Let's try again:

5. In 2014, the Spurs won *z* games, and the Rockets won 4 more games than the Spurs. In 2015, the Rockets won twice as many games as they did in 2014. In terms of *z* how many games did the Rockets win in 2015?

Great. Note that mathematically, question 5 is *exactly* the same as question 1. The only difference is the letter used for the variables and the concepts they represent. But do you think the question will be perceived the same way by your students? Which version do you think is easier or harder? Or are they both the same? I would argue that question 5 is probably harder than question 1 because of the increased abstraction of the concepts.

As math teachers, it can sometimes be tempting to treat questions with identical underlying math as identical problems, but students don't always see it that way. We've all seen students who can perform well on pure equations (which can often be done through rote memorization of processes and muscle memory) but perform poorly when the same concepts appear in a context.

This is a particularly important issue if you're teaching non-native English speakers. ELL students may be excellent math students but struggle to decipher long English passages. It can even be an issue for students who are native speakers but are just poor readers. For them, the ACT is twice as much work. And twice as much work for you, too. First you have to teach them the underlying math concepts they need to get the question. Then you have to teach them how to identify those concepts based on the question as it's written.

Similarly, sometimes students respond differently to different kinds of contexts. They might have no problem with questions that are about everyday household events but struggle with questions about science. But as we've seen in our translations above, problems can have the exact same underlying math but be put in different contexts. To wit, compare this problem:

6. **Sherwyn deposits $10,000 in a bank account that gets 5% annual compound interest. Which of the following equations describes the amount of money, *C*, Sherwyn will have in his account after *t* years?**

. . . with this problem:

7. **Sherwyn determines that a certain region contains 10,000 frogs. If the frog population grows by 5% each year, which of the following equations describes the number of frogs, *F*, in the region after *t* years?**

These problems contain exactly the same mathematical principles, but one is in an economic context and the other in a scientific context. We could even remove the context altogether:

8. **Which of the following equations shows an exponential function with 5% growth and an initial value of 10,000?**

Is it possible that some students are better at one context than another? Sure, absolutely. But does that tell us anything meaningful about those students' mathematical abilities?

These issues teach us a few important lessons about context:

- Context is irrelevant to mathematical content. The same content can be provided in different contexts.
- Context can affect the difficulty of a question, in particular by making things seem more abstract or concrete.
- Some students will struggle with certain contexts due to individual preferences or skills. Some will struggle with all contexts due to poor reading comprehension.
- Students' performance in different contexts may be significant but tells us little about their understanding of the underlying mathematical principles.
- We should strive to get students to see the *irrelevance of context*, to ignore context and focus on the math.
- We can demonstrate the malleability of context by taking any problem and putting it in a new context or stripping the context away altogether.

Distractor Choices

Let's go back to our first sample question:

1. **Bob has 4 dollars more than Lisa. If Lisa has x dollars, how much would Bob have if he doubled his money?**
   A. $x + 4$
   B. $x + 8$
   C. $2x$
   D. $2x + 4$
   E. $2x + 8$

How do we do this question? We want to take these sentences in English and rewrite them as algebraic expressions.

- Lisa has x dollars.
- Bob has 4 more, so Bob has $x + 4$ dollars
- If Bob doubled his money he'd have $2(x + 4)$
- Distribute across the parentheses to get $2x + 8$. That's choice E.

Let's look at the wrong answers here. What do you think the most popular wrong answer would be? It's probably choice D, $2x + 4$. You get that if you forget to distribute the 2 inside the parentheses. This is particularly common if you don't write down your work and only think about it in your head, since we don't tend to pronounce parentheses:

> *Bob has x plus four, so if he doubled his money he'd have two times x plus four. Ah, choice D says two times x plus four. I am finished.*

What about the other choices? Choice A, $x + 4$, is the amount Bob has now, before he doubles his money. Choice B, $x + 8$, has the opposite problem as choice D: You doubled the 4 but not the x.

We call wrong answers *distractor choices*. Strictly speaking, the term "distractor" refers to any wrong answer choice in a problem, since they distract you from the right answer. But some wrong choices are more distracting than others. The ACT loves to plant incorrect choices that aren't random but are answers students would get by making common mistakes. This seems evil to students, and, well, it's not not-evil, but it is important information for you because it makes it easier for you to diagnose their mistakes. If you know a student picked D on this question, you know exactly what they did wrong before they even say anything. A-List offers an online remote test-grading service that provides reports with answer-choice level information for your class. These reports show the percentage of students who chose each choice for each question. This way, you can immediately see not only which questions most students got wrong, but which wrong answers were most popular.

Not every choice will be equally tempting of course. Some wrong answers really are just random numbers that only students who pick randomly will choose. Others are so tempting that their mere presence is the only thing that makes a hard question hard: The underlying math is easy, but *everybody* looks at D and picks it right away, without thinking.

You should explicitly talk about distractor choices when reviewing real tests, especially if you see one that's particularly tempting. Don't just give the right answer and stop. Look at the wrongs and ask: Why might someone think it's A? Why is A not the answer?

When these ideas are out in the open, students will be prepared to face them.

When you're translating a question, you'll want to retain the distractors that are already there, but you may have to adapt them for any new numbers you picked. For example, let's look at the new version of question 1 that we came up with, now with answer choices.

5. In 2014, the Spurs won z games, and the Rockets won 4 more games than the Spurs. In 2015, the Rockets won twice as many games as they did in 2014. In terms of z how many games did the Rockets win in 2015?

A. $z + 4$
B. $z + 8$
C. $2z$
D. $2z + 4$
E. $2z + 8$

Here we have all the same numbers as the original, so our choices should be exactly the same, just using z instead of x. But let's say we changed the 4 to a 10. We'd change the correct answer to $2x + 20$, of course, but we'd also want to change the wrong answers analogously:

9. In 2014, the Spurs won z games, and the Rockets won 10 more games than the Spurs. In 2015, the Rockets won twice as many games as they did in 2014. In terms of z how many games did the Rockets win in 2015?

A. $z + 10$
B. $z + 20$
C. $2z$
D. $2z + 10$
E. $2z + 20$

Now let's stray further from the original. Let's make the Rockets worse in 2015:

10. In 2014, the Spurs won z games, and the Rockets won 10 more games than the Spurs. In 2015, the Rockets won *half* as many games as they did in 2014. In terms of z how many games did the Rockets win in 2015?

Now instead of 2(z + 10), the correct answer should be $\frac{z+10}{2}$. Do we want the answer to appear like that? Or $\frac{z}{2}+5$? Or $\frac{1}{2}z+5$? Let's go with the last option to make it most analogous to the original. We can fill out the rest of the choices the same way:

10. In 2014, the Spurs won z games, and the Rockets won 10 more games than the Spurs. In 2015, the Rockets won *half* as many games as they did in 2014. In terms of z how many games did the Rockets win in 2015?

A. $z + 5$

B. $z + 10$

C. $\frac{1}{2}z$

D. $\frac{1}{2}z + 5$

E. $\frac{1}{2}z + 10$

First we should note that the answer is now D instead of E. Choices are generally arranged in order from smallest to largest when they give values. When the choices are expressions like this there's a bit more leeway about the order, but they're still grouped in some logical way.

But now we should think: Maybe in this new version students won't make exactly the same mistakes. In the original, kids might mess up how to apply the 2, but they'll all correctly *multiply* the two. Fractions confuse students, and it's certainly possible students will get confused about which direction to go after they get to z + 10. So let's add a distractor where they multiply by two instead of dividing by two:

11. In 2014, the Spurs won z games, and the Rockets won 10 more games than the Spurs. In 2015, the Rockets won *half* as many games as they did in 2014. In terms of z how many games did the Rockets win in 2015?

A. $\frac{1}{2}z + 5$

B. $\frac{1}{2}z + 10$

C. $z + 5$
D. $z + 10$
E. $2z + 20$

Adapted Translation

We've seen ways for you to alter existing problems while retaining the original concepts in the problem. This is useful for several reasons:

- To demonstrate the irrelevance of context to the underlying math
- To create more problems in particular contexts that are difficult or troublesome for your students
- To create more problems dealing with a specific concept

But who says you have to stick to the original concepts? We've already seen how we can adapt problems beyond their original scope. While we're at it, let's keep going and add more concepts.

Let's start with our original problem:

1. Bob has 4 dollars more than Lisa. If Lisa has x dollars, how much would Bob have if he doubled his money?
A. $x + 4$
B. $x + 8$
C. $2x$
D. $2x + 4$
E. $2x + 8$

First, let's try to make it *easier*. Let's change it from an algebra problem to an arithmetic problem:

12. In 2014, the Rockets won 10 more games than the Spurs. In 2015, the Rockets won half as many games as they did in 2014. If the Spurs won 40 games in 2014, how many games did the Rockets win in 2015?

The Rockets won 10 + 40 in 2014—that's 50 games. They won half as many in 2015—that's 25 games.

Question 12 is now easier than question 11 because we're dealing with concrete numbers. There's no danger of misapplying the parentheses because there are no parentheses.

This, by the way, is why we advocate a technique called Plug In for questions like number 11. Pick a number for z to get an answer for the number of games. Then put your number in the choices for z and see which gives you the same answer. You can do this for just about any question that has variables in the answer choices. If the answer is $x + 4$, it's *always* $x + 4$, regardless of what the value of x is. So pick one and try it.

If the problem is too easy for you now, we can keep it arithmetic but make it harder by using harder numbers:

13. In 2014, the Rockets won 5 fewer games than the Spurs. In 2015, the Rockets won 7/8 as many games as they did in 2014. If the Spurs won 61 games in 2014, how many games did the Rockets win in 2015?

Ugh. Okay, fine. 61 – 5 = 56 wins in 2014. (7/8) × 56 = 49 wins in 2015.

That one was still too easy because we had to stick to integers because we're talking about games. Let's try going back to the original version but make the numbers uglier:

14. Bob has $3.80 more than Lisa. If Lisa has $2.18, how much would Bob have if he doubled his money?

Now it's harder, sure. But the difficulty here is purely computational. If students have calculators, it's a snap. Question 13 is probably a bit harder because students have to deal with the fraction, 7/8.

Okay, let's go back to algebra, but we'll add in a new concept to make it harder. What's something that students struggle with? How about percent change?

15. Bob's salary is 30% greater than Lisa's.

Wait, stop, no. Let's not make the man's salary higher than the woman's. Start again:

15. Molly's salary is 30% greater than Noah's. If Noah's salary is x dollars, which of the following gives Molly's salary after she gets a 20% raise?

Our previous translations were variations on the same problem: The numbers were different but the concepts were the same. This problem now significantly different than the original because, it's introduced a new concept of percent change.

Let's do the problem first:

- Noah gets x dollars.
- Molly gets 30% more, so she gets $1.3x$.
- She gets a 15% raise over what she had, so that's $1.2(1.3x)$.
- Distribute across the parentheses to get $\mathbf{1.56x}$.

Let's call that choice A. We could leave it as a decimal like that. Or use $\left(\frac{156}{100}\right)x$. Sometimes ACT problems may be written to make the math steps apparent, so you could make the choice $(1.2)(1.3)x$, or even $x + 0.56x$.

Here are some possible distractor choices:

B. $0.06x$ That's just 30% times 20%.
C. $0.26x$ That's 20% of 130%.
D. $0.36x$ That's 30% of 120%.
E. $1.3x$ That's her starting salary.
F. $1.5x$ That's adding 30% and 20%.

All of these distractors assume that students know to multiply percentages. But maybe they don't! We could have distractors like these:

G. $30x + 20$

H. $\frac{x}{1.56}$

J. $\frac{1.3x}{1.2}$

Now we're just shuffling these numbers around randomly, but such things can still be tempting for students who have no idea how to proceed. In general, the more tempting the distractor choices, the harder the question is. Some students may get choice C above just because they're careless. If a student gets that as an answer and sees that it's a choice, they'll pick it and move on without thinking. But if it's *not* a choice, they'll know they made a mistake and are more likely to go back and check their work. So having *random* distractors doesn't guarantee that more students will *understand* a question, but it makes it less likely they'll make *careless* mistakes.

Let's see a variation on this question in order to demonstrate that.

16. Molly's salary is 30% greater than Noah's. If Molly gets a 20% raise, her salary is now what percent greater than Noah's?

A. 6%
B. 10%
C. 50%
D. 56%
E. 60%

Every single student is going to pick choice C here. Every single one. It is so tempting to just add 30 and 20 to get 50. So when they see that 50 there, their instincts are confirmed and they move on to the next problem. But if 50 wasn't there, if instead our choices were:

A. 26%
B. 36%
C. 46%
D. 56%
E. 66%

then they'd know something was up. That doesn't mean they'd necessarily know the *right* way to do the problem, but they'd know their instinct was wrong. They'd spend more time thinking about it, and on average, more students would get it right.

Let's keep adding concepts. Let's add a second equation here.

16. If Molly's salary is 30% greater than Noah's and their combined salary is $92,000, what is the value of Molly's salary?

Now we have two equations

$m = 1.3n$
$m + n = 92{,}000$

We can substitute to get

$1.3n + n = 92{,}000$
$2.3n = 92{,}000$
$n = 40{,}000$

That's Noah's salary. Subtract from 92,000 to get

$m = 52{,}000$

Let's look at those numbers we used up to get our answer. Do any of those look like good numbers to use as a distractor choice? YES. 40,000! *You should absolutely include 40,000 as a distractor choice.* If you do, half your students will pick it. They will go through all the correct steps, get a value for one of the variables, then assume that must be the answer without actually checking if that's the variable they want.

This is a great method for coming up with distractor choices. Go through all the correct math you've done and look for numbers you used that weren't the correct answer. Use those as choices. Somebody will get that number and stop.

Another great distractor idea: Just take 30% of $92,000. That's $27,600. Great! Now subtract that from $92,000. That's $64,400. Perfect! We've got three distractor answer choices now. Throw in one more number for a fifth choice. It could be anything—they don't all have to be tempting—just make sure it's on a similar scale to the others (don't choose $7).

16. If Molly's salary is 30% greater than Noah's and their combined salary is $92,000, what is the value of Molly's salary?
A. $27,600
B. $35,000
C. $40,000
D. $52,000
E. $64,400

Now that we have answer choices, this question is a great demonstration of our other math technique Backsolve. Pick an answer choice and put it into the problem to see if the numbers work out. Try choice C, $40,000. If that's the answer, then that's Molly's salary. So $40,000 is 30% higher than Noah's. Divide by 1.3 to get Noah's salary is $30,769.23

If this is right, the two salaries will add up to $92,000. They obviously don't; they only add up to around 70,000. So choice C was too small. Try the same with choice D and it will work out.

The nice thing about Backsolve is you start directly from the question so you're much less likely to choose the wrong variable. The whole first step is, let's make the choice the answer to the question. You still need to know how to manipulate percentages, but you're less likely to make a careless error.

We can see now that adapted translation is a great way to make your own questions when you want drills on specific concepts. Want to work on systems of equations? Find questions with one equation and add another parameter. Want to work on graphing lines? Find questions with ungraphed linear functions and ask students to graph them. Do both of these things together: Add an equation and graph them both. Look: We started with one fairly easy problem and made 16 versions of it. And we didn't even add that many concepts to it. The only limits are your imagination.

Student Work

That seems great and all, but that seems like a lot of work. How can we get drills to the students faster?

Why not have the students themselves do the work?

This may seem like a cheap way for you to get out of work, but it isn't (entirely). The discussion above shows that even simple problems can spur long discussions of what is being tested and how to go about solving problems. These are conversations that we can be having with our students explicitly.

The issue of distractor choices, for example, shouldn't be a secret. It's important that they know about them, that the people who write the questions *intentionally* make wrong answers available that are the result of predictable errors.

1. **Pick a problem**. Or, assign problems to the students.
2. **Translate the problem** into a new problem. You can give specific parameters for what the problem looks like. For example:

   a. Be sure the translated problem contains none of the same numbers as the original. (Or set limits on how many numbers they're allowed to keep.)
   b. Be sure the translated problem is in a different context. Don't just change basketball teams to baseball teams; make it significantly different. You can specify the context if you like (e.g., make it a science question).
3. **Write distractor choices**. Be sure at least two incorrect choices have specific incorrect paths that lead toward them.
4. **Write answer explanations for the question**. This is a great tool. Nothing gets students to dive into a problem like being forced to explain it to another student. Be sure to:
   a. Include two ways of solving the problem. They don't have to be equally good ways—one might be clearly faster or easier than the other—but they both have to be mathematically valid ways.
   b. Explain why students might pick the distractor choices. What mistakes would lead to those choices.
5. **Analyze how the problem has changed**. Has the translation made the problem easier or harder? Why?

In reviewing the students' work, there will likely be some translations that don't work. The original problem was fine but the translation is problematic or outright nonsensical. See, for example, our awkward wording for question 4 above. If it doesn't work, talk about it. *Why* didn't it work? What did the student do to the problem that made it problematic?

Be sure also to check that the students' wrong answers are all actually wrong! Sometimes students will accidentally make a wrong answer right in their efforts to be sneaky. For example, they might be given two choices with algebraic expressions that seem distinct but are actually equivalent, like $2x + 8$ and $2(x + 4)$.

This is also a great exercise to do in groups. Have every student write a translation, then pass it to the left so another writes the distractor choices. Then pass it down once more to have a third write the answer explanations. When they're done, they'll have their own drill. Have groups swap drills and do each other's questions. See how they do!

Translation Summary

- Pick an ACT problem. Any problem.
  - Analyze the concepts involved in the problem. What rules, skills do you need here?
  - Analyze the answer choices. Which ones are tempting? Why might a student pick one?
- Write a new problem using the *same* skills in a new context.
  - Try writing the same problem in other settings. Try science or history/social studies. Try writing it with no setting at all.
  - Did your problem come out easier, harder, or the same? Why?
  - Change the numbers to make it easier. Change the numbers to make it harder (but without changing the math content).
  - Write new answer choices. Preserve any distractor choices from the original, adapted for your new numbers. Try coming up with new distractor choices.
- Keep the same framework as the original, but *add concepts* to the problem. For example, if it was one equation translation, make it a system. Add percentages. Add the coordinate plane.
  - How would that alter the difficulty?
  - Could you use a math technique on the original problem? Can you use one on your new problem? Try to adapt the original into a Plug In or a Backsolve problem.
  - Adapt distractor choices to fit the new content. Find distractors by looking at the math work you did in order to get the right answer. Use some of your intermediary steps as wrong choices.

4

ACT Science in the Classroom

ACT Science can be a tough sell for schools. It doesn't neatly fall into your existing curriculum the way that math does. ACT Math is math: It tests triangles; you already talk about triangles in class. But ACT Science isn't quite the same as school science. You'll see familiar concepts, sure—circuit diagrams, titration, kinetic energy, and all your favorites—but it's not like an AP test. AP tests require you to know a lot of facts. The ACT is more about the act of being a scientist. How well can you read data representations and analyze the figures and relationships described? How well can you form conclusions based on those data? How well can you understand why experiments were set up as they were? In other words, ACT Science is not testing whether you *know science*; it's testing your ability to *do science*.

Practice Sections

Everything we already said about math practice applies equally to the Science Test. It's important to get real practice. All the sources we listed for practice math material can also be used to find practice science material.

It's still best to do practice in full proctored test conditions. You can often see why this is true most clearly on the Science Test because it's *last*. Students who just do a single Science section in isolation tend to do better than those who do it after having spent two and a half hours doing English, math, and reading. Especially reading; that one really drains them. Doing a science passage after having read four long confusing reading passages can be so exhausting. But that's the way it will be on the real test, so it's best to get used to it if you can.

But if full proctored tests aren't possible with your schedule, doing a timed Science section in class is pretty easy since it's only 35 minutes long. And whether you do sections in class or in a proctored environment, you'll want to review those sections in class after the fact.

Doing existing ACT practice sections is fine, but how do we go beyond that? What other activities can we do to help with this section? That partly depends on what class you're teaching.

ACT Science College and Career Readiness Standards are grouped into three strands:

1. **Interpretation of Data**. Read and understand data representations (tables, graphs, diagrams, etc.).
2. **Scientific Investigation**. Understand why and how experiments are set up as they are.
3. **Evaluation of Models, Inferences, and Experimental Results**. Draw conclusions about results and evaluate hypotheses.

These strands broadly correspond to the three types of passages—Data Representation, Research Summaries, and Conflicting Viewpoints—but they certainly overlap.

Strands 2 and 3 are directly related to science, so they are well-suited to be discussed in science class. Strand 1 overlaps quite a bit with math content so it is well-suited to be discussed in math classes, so it's a good topic to broach there. But all three of these strands can easily be addressed in non-STEM classes as well. Social sciences like economics, sociology, or history often use tables, graphs, and data representations. They must set up studies and model expected results. They must analyze data and draw conclusions based on results. All of these are concepts that are central to the ACT Science test.

For the purposes of this book, we'll focus on how to incorporate ACT Science material into math and science classes, but there are opportunities to include this material in a wide range of classes.

Science Classes

Science Knowledge

Again, don't worry too much about knowledge of science facts. First of all, science teachers are already doing this all day long. You don't need us to tell to teach science. You've got a lot of ground to cover, don't let us get in your way.

But also remember that knowledge of science facts is a small part of the ACT. There will certainly be a lot of science facts on the test. There's *tons* of science on the test. And not kid science either: real, fancy science. The following table shows the subjects of the passages from the first practice test in *The Official ACT Prep Guide 2016–2017*. As you can see, there's a wide range of sophisticated topics.

Passage	Type	Subject	Content
1	Data Representation	Biology	Fur pigmentation in deer mice
2	Research Summary	Chemistry	Filtration of dissolved nickel (Ni^{2+})
3	Conflicting Viewpoints	Physics	Gravity and radiation pressure in star formation
4	Research Summary	Earth Science	Vermicompost and plant yield
5	Research Summary	Physics	Cathode ray tubes and positioning of light spots
6	Data Representation	Chemistry	Mean free path of noble gas atoms

But for all of these passages, you don't need any *existing* knowledge. The passages define terms for you, provide necessary background information, and describe experiment set-up in detail. You needn't have ever even heard the term "vermicompost" to be able to do all the questions in passage 4. This is one of the reasons that the Science Test is more analogous to the Reading Test. On the Reading Test, you don't have to know anything about Thomas Jefferson to do all the questions on a passage about Thomas Jefferson, because the passage tells you everything you need. The same is true on the Science Test. Everything you need to know is given to you.

That's not to say there will be *no* need for science facts. There will certainly be some questions that may explicitly require you to know the meanings of terms that are not defined in the passage, such as density, gravitational potential energy, or the law of conservation of mass. But these questions are a small percentage of the total questions, usually around 5% or two questions per test. Cramming with flashcards might help for AP tests, but it's a waste of time for the ACT. You won't need to memorize the Periodic Table, and you won't need to know all the steps of the Krebs Cycle.

That's not to say that having science knowledge is useless. On the contrary, having a solid background in a topic can make you more comfortable with a passage on that topic. And having a general understanding of a concept can often be crucial, even if the particulars are given to you in the passage. For example, a passage might describe some experiments that demonstrate a given chemical reaction. You likely don't need to know anything about the *particular* reaction nor the chemicals involved. But you probably should know what a chemical reaction *is*: How they work and how to read an equation. It's probably *technically* possible to do the passage without knowing anything about reactions, but it's likely much harder.

(On the other hand sometimes too much knowledge gets in the way. Students spend mental energy trying to remember things they've learned that are only tangentially relevant to the experiment, when they should just be looking up data in the table.)

This is an important point that should be brought up frequently. A lot of students are scared of science and big science words. They feel inadequate about the things they don't know or only partially know. Yet most of the time students don't need anything at all.

To demonstrate this, we often give sample passages that contain *completely fictitious science*. Here's an example from our textbook, *The Book of Knowledge*:

Passage I

Certain organic substances can undergo a process called *mimeticization*. The *cromulence* of a substance determines the rate at which it becomes mimeticized. Cromulence is measured in peccaries (pec). The *grittiness* of a substance is the factor by which the substance can be syllogized when mimeticization is taking place. Grittiness is measured in ecksteins (eck). Table 1 identifies the cromulence and grittiness of several liquids at 20°C.

Substance	Density (g/cm^3)	Cromulence (pec)	Grittiness (eck)
pentane	0.626	16.54	0.102
hexane	0.659	30.84	0.772
heptane	0.684	46.19	1.442
octane	0.703	60.41	2.121
nonane	0.718	73.99	2.873

Table 1

Obviously, this is all made-up nonsense. We threw in some real organic compounds to give it the luster of legitimacy, but that's just a mirage. These are fake concepts. There's no such thing as the cromulence of mimeticization. But it doesn't matter if they're fake because ACT Science passages only ask you to read and understand the information provided. That information is enough to do a question like this one:

3. In order to synthesize a certain type of stigmagated plastic, engineers need a mimeticized chemical that has a grittiness between 2.0 eck and 2.5 eck. Based on the information in Table 1, which of the following substances would be suitable for making this plastic?
A. Pentane
B. Hexane
C. Heptane
D. Octane

Making stuff up can be fun, but it's not just a game. It helps demystify the fancy science words. A life in science always involves reading about new concepts that you're seeing for the first time. Getting students to learn to deal with these concepts rather than freak out about them is a big step forward.

Experiment Design

So if you shouldn't focus on science facts, what should you do?

A great area to focus on is *Scientific Investigation,* the second strand in the ACT Science CCRS. The strand roughly corresponds to Research Summaries passages, those which contain several different experiments about a single concept. These passages make up the largest chunk of the test, comprising about half of all the questions.

One of the more common question types in these passages is what we call *Experiment Design Questions*. These are questions that specifically ask about the way the experiments were set up and conducted. These can come in a range of topics, all of which overlap. For example:

- *How was the experiment set up?* Some questions simply ask you to understand the procedure. Where was the test tube? What does that valve do? These questions on a basic level simply ask what *happened,* and they often require little more than reading the accompanying text. Harder questions may involve more complex experiments that sometimes come with diagrams and pictures.
- *Why was the experiment set up that way?* Why did we close that valve? Why did we measure the weight of the flask before adding the solution? Why did we heat the solution after adding the powder? What would have happened if we did things differently? These questions require a deeper understanding of how to set up experiments and what they teach us.
- *What variables were tested?* Often passages will have several experiments that have different variables controlled and varied. Experiment 1 will keep mass and temperature constant but vary the chemicals used, while Experiment 2 will keep mass and chemicals constant but vary the temperature. On simpler questions, students may be asked to identify which variables were varied or kept constant.
- *Why were those variables tested?* Harder questions might ask *why* certain variables were varied or kept constant, and what that means about the conclusions we can draw. This gets into questions about the purpose of the experiment. Independent of what the results actually said, what did we hope to learn in this experiment?

All of these questions revolve around the deepest bases of the scientific method and get at the fundamental ideas of what it means to do science. If you do anything in your science class toward preparing for the ACT, spend some time talking about experiment design.

Of course, if you've got any kind of lab to go with your lectures, there's a good chance you're already talking about experiment design. Great! Keep it up. Lab work is the most productive way you can spend time for the Science Test.

But you can go further and frame these conversations explicitly around the test itself. Take an experiment you're already doing (or planning to do) give a little quiz asking some the above questions about it. Make sure they understand the variables that were controlled and varied. Make sure they understand the purpose behind the different parts. You can even give the quiz before running the experiment; all these questions are independent of the actual results.

A good group exercise is to have students come up with their own experiment ideas. This is in essence no different than normal Science Fair shtick. The difference is that since we're just talking about design in the abstract, they don't have to be experiments that students actually perform. That's good because it means expense is no object. They can be thought experiments about how we *would* go about testing a question without having to go through the actual grunt work.

Students who struggle with these concepts can start with simple ideas. Does a certain fertilizer promote plant growth? How would you test that? How would you physically set up the experiment? What variables would be controlled and which would be varied? Concepts like independent and dependent variables are fundamental to understanding how experiments work, indeed how all of science works. Again, since we're not actually conducting experiments we can move on to more elaborate projects. Does this toothpaste whiten teeth? Does this food product have the health benefits it claims to have? You can make it more relevant by using whatever topics you've already been talking about in your class as a starting point, or just your imagination.

Framing these as in-class discussions can allow you to play devil's advocate in their experiment design. Talk to them about potential sources of error in their measurement. Bring up other factors that could introduce bias their results. How can they change their experiments to correct for those biases? These are issues that come up explicitly on ACT questions.

You can also have students work in small groups to come up with ACT-style quizzes for each other. Have one group design an experiment and write a passage describing it along with simple questions about the set up and the variables. Then have the groups switch and take each others' quizzes.

Evaluation of Models, Inferences, and Experimental Results

All these topics lead to the ACT Science Standards' strand 3, *Evaluation of Models, Inferences, and Experimental Results.* These are questions that ask students to actually draw conclusions from the data. In some respects, these questions have a lot of overlap with the issues we discussed in experiment design. They may require you to know the reasoning behind an experiment and understand how the variables interact with each other. On the other hand, such questions generally require actual data, not just the framework experiment. You can't draw conclusions from results if you don't have any results.

You can still have these conversations when discussing hypothetical experiments by also talking about hypothetical results. When you're setting up an experiment, it's best to have some sort of hypothesis ready that you experiment will test, e.g., will this fertilizer make plants grow taller? You then also must have a vision of what it would mean for your experiment to confirm your hypothesis. How would you know if you're right? What kind of results would I expect? Or, from the other direction, if I got a particular set of results, what would that mean about my hypothesis?

Math Classes

Schools who do SAT preparation in their classes often split the test sections, so ELA teachers cover the English and Reading Tests, and math teachers get the Math and Science Tests. That means if you're a math teacher you may find yourself talking about the Science Test in your math class. Great!

Interpretation of Data

As we said, you don't have to worry about science knowledge, so don't bother dusting off your old physics textbook. Note that even though you'll be dealing with a lot of numbers on these figures, you rarely have to do any computation. Most of the questions are simply looking up values, getting ranges of values, or seeing relationships between values. Occasionally, you may have to do some light arithmetic, but only the sort that you can easily do by hand or via approximation. You are not permitted to use a calculator, but you should not need to.

So why are you talking about this test in math class at all? Because of strand 1 from the ACT Science standards: Interpretation of Data.

This strand roughly corresponds to Data Representation passages, which present a single set of graphs and figures (as opposed to multiple different experiments in the Research Summaries passages). Most of the questions in these passages boil down simply to understanding the figures and knowing how to extract data from them. But these questions aren't isolated to the Data Representation Passages; Research Summaries passages have lots of data questions, too. Even Conflicting Viewpoints passages, which are usually primarily text based, will occasionally have some figures with a data question or two. All in all, over 60% of the questions directly involve reading and interpreting the figures.

What Kind of Data?

As we mentioned, you'll want to get your students some real ACT Science sections at some point, but that only gets you so far. You'll want to supplement that with some material of your own. It can be hard to write a Science passage, though, especially if you're not a scientist.

But it doesn't have to be about science! Remember that science knowledge isn't important here. All we're trying to do is teach students to read tables and graphs. As such, *any* tables and graphs will do. Besides scientific topics, they could show demographic trends, sociological studies, public transit usage, TV ratings, movie box-office earnings, *literally anything*.

Remember: As we saw earlier with the cromulence of mimeticization, the passage doesn't even have to be *true*. It just has to be clear and consistent. Just be sure that all the necessary concepts are clearly defined such that students don't need any prior knowledge to answer your questions.

Where Can I Get Data?

Outside of actual ACT sections, there are lots of places you can find tables and figures to practice on. There's data everywhere these days.

First stop: the Internet. Hoo-boy does the Internet love data and data visualizations. You can start with sites that specialize in data, fivethirtyeight.com most notable among them. Specialized scientific sites like *Discover* are also often fruitful sources. Sports sites are often fertile ground for data, although such data can often be very specialized, and you don't want to alienate students who may not follow the sport in question. But even regular newspaper articles are often supplemented by tables and graphs. Open up today's *New York Times* or *Wall Street Journal* and you're likely to see some tables or graphs.

If you want some ready-made questions, you can even sneak a peek at the SAT. While the SAT doesn't have an analogue to the ACT Science test, the redesign of the SAT in 2016 has led to an increase in the number of questions involving data interpretation. It's not just the math: Figures have been added to passages in the Reading and Writing sections, too.

You could also create your own tables and graphs. This is a nice opportunity to choose data on whatever topic interests you. You like birdwatching? Make tables about bird species distribution. Like baseball? Get some baseball cards. (Man-oh-man, do baseball nerds love to make graphs.) Like *X-Men*? We wrote a science passage about the genetics of mutants in *The Book of Knowledge*. And of course, if you're making your own tables and graphs anyway, you could make them entirely fictitious. The data doesn't have to be *true*, only clear and consistent.

Data Questions

Once you have some figures, you'll also want some questions to go with them. We already mentioned some common question types in Part 1, but here's a refresher:

- **Data Lookup Questions**. These are the most common type of question on the Science Test and also the simplest. They require students to retrieve a value given in one of the tables or graphs provided. Often, students simply must match the name of the quantity mentioned in the question with the headings and labels listed in the figure. If students aren't getting near 100% of these questions, this skill should be drilled into them.
- **Combination Questions**. These questions are often similar to Data Lookup Questions, but they require using multiple tables or graphs to find the needed data. Table 1 compares A to B and Table 2 compares B to C, so you must combine them to compare A to C.
- **Relationship Questions**. Instead of asking to retrieve specific data values, these questions ask about the relationship between fields. Usually this is just a matter of figuring out up vs. down: When one value goes up, does the other value always go up, always go down, do both, or stay the same? Note that we don't need to go so far as to show causation, just show trends of movement.
- **Inferred Data Questions**. These questions are similar to Data Lookup Questions with one important twist: The data point that the question is asking for does not literally appear in the figure. Students must infer the value based on the data that do appear. Often these inferences don't have to be exact, just in a range. I know I want something higher than point A, which was 5, and lower than point B, which was 10, so my answer will be between 5 and 10.

The hard part here is finding the data. Once you've got a couple of tables and graphs—and really even just one figure is enough—it's very easy to throw on some questions like these. Data Representation passages generally have five or six questions, so you can make it a short quiz. If it seems like your questions are too easy, you can always obfuscate by throwing in some tough language or extra information. Remember this question:

> **3. In order to synthesize a certain type of stigmagated plastic, engineers need a mimeticized chemical that has a grittiness between 2.0 eck and 2.5 eck. Based on the information in Table 1, which of the following substances would be suitable for making this plastic?**

Really all it's asking is:

> **3. Which substance has a grittiness between 2.0 and 2.5?**

We can see that phrasing matters.

Reasoning Questions

This is another opportunity to discuss the ACT Standards' Strand 3, *Evaluation of Models, Inferences, and Experimental Results.* We mentioned this strand in the section on science classes, saying you can pivot from conversations about understanding the purposes behind the design of an experiment to visualizing what it would mean for a hypothesis to be confirmed.

You can also talk about such issues once you actually have some data in a figure. You don't need to have hypotheses baked into the figures; you can insert claims to be tested in the questions themselves. Introduce a claim in the question text that makes a specific claim or prediction about your data, then ask whether your data backs this up. You can then include choices that each first say yes or no, and then give a reason why the data does or does not support the claim. These are the *Reasoning Questions* we discussed with respect to standard R.CCR.8 in Part 2 above.

Have Students Make the Drills

We've just talked about how easy it is to find passages and write questions—so why not have the kids do it themselves? Besides being less work for all of us (whoo!), getting them to see the principles behind the construction of the test can help them better understand what the test is asking them to do.

This can be a fun group project as well:

- First, have them make some figures. You can assign rules about what kind of figures, if you like, for example, one table and one line graph. They can use any source they like: their own research, real data from science labs, or completely made up. You can let them have fun with this and choose less serious texts to use as passages.
- Then, have them write five questions. You can give specific assignments about what question types they must use, for example, at least three different types and two questions must be data lookup, etc.
- Have them write answer choices, too, including some tempting distractor choices for kids who look up the wrong column.
- Have them type everything up along with sufficient explanatory text such that all terms are clearly defined.
- Finally, have the groups switch so that each group takes a different group's quiz.

Appendix
All Math Alignment Tables

How to Read the Tables

Each table refers to a single conceptual category. Within each table, standards are grouped by domain and cluster. Remember that standards are numbered continuously within each domain, irrespective of clusters. Each table contains columns showing:

- The code for the standard, as defined by the CCSSI
- The standard itself
- Its alignment with the ACT

Each standard has a code containing three parts: a letter denoting the conceptual category, a series of letters denoting the domain, and a number denoting the standard. If the standard has any subskills, the skills will be denoted by a letter following the standard number. **Code: [category]-[domain].[standard][subskill]**

For example, "F-IF.7c" denotes the Functions category [F], the "Interpreting Functions" domain [IF], the seventh standard within that domain [7], and the third subskill associated with that standard [c].

The two alignment columns will each display one of the following symbols:

- **Y** = The standard is aligned with the ACT.
- **N** = The standard is not aligned with the ACT.
- **P** = The standard is partially aligned with ACT. This means certain words or phrases were excluded in the alignment document.

Standards marked with * are Modeling standards. Standards marked with (+) are standards beyond College and Career Readiness.

Code	Standards for Mathematical Practice [MP]	Aligns
MP.1	1. Make sense of problems and persevere in solving them.	**Y**
MP.2	2. Reason abstractly and quantitatively.	**Y**
MP.3	3. Construct viable arguments and critique the reasoning of others.	**Y**
MP.4	4. Model with mathematics.	**Y**
MP.5	5. Use appropriate tools strategically.	**N**
MP.6	6. Attend to precision.	**Y**
MP.7	7. Look for and make use of structure.	**Y**
MP.8	8. Look for and express regularity in repeated reasoning.	**Y**

Code	Standard	Aligns
Number and Quantity		
The Real Number System [N-RN]		
Extend the properties of exponents to rational exponents.		
N-RN.1	1. Explain how the definition of the meaning of rational exponents follows from extending the properties of integer exponents to those values, allowing for a notation for radicals in terms of rational exponents. *For example, we define* $5^{1/3}$ *to be the cube root of 5 because we want* $(5^{1/3})^3 = 5^{(1/3)3}$ *to hold, so* $(5^{1/3})^3$ *must equal 5.*	**Y**
N-RN.2	2. Rewrite expressions involving radicals and rational exponents using the properties of exponents.	**Y**
Use properties of rational and irrational numbers.		
N-RN.3	3. Explain why the sum or product of two rational numbers is rational; that the sum of a rational number and an irrational number is irrational; and that the product of a nonzero rational number and an irrational number is irrational.	**Y**
Quantities* [N-Q]		
Reason quantitatively and use units to solve problems.		
N-Q.1	1. Use units as a way to understand problems and to guide the solution of multi-step problems; choose and interpret units consistently in formulas; choose and interpret the scale and the origin in graphs and data displays.	**Y**
N-Q.2	2. Define appropriate quantities for the purpose of descriptive modeling.	**Y**
N-Q.3	3. Choose a level of accuracy appropriate to limitations on measurement when reporting quantities.	**Y**
The Complex Number System [N-CN]		
Perform arithmetic operations with complex numbers.		
N-CN.1	1. Know there is a complex number *i* such that $i^2 = -1$, and every complex number has the form $a + bi$ with a and b real.	**Y**
N-CN.2	2. Use the relation $i^2 = -1$ and the commutative, associative, and distributive properties to add, subtract, and multiply complex numbers.	**Y**
N-CN.3	3. (+) Find the conjugate of a complex number; use conjugates to find moduli and quotients of complex numbers.	**Y**

(Continued)

Code	Standard	Aligns
Represent complex numbers and their operations on the complex plane.		
N-CN.4	4. (+) Represent complex numbers on the complex plane in rectangular and polar form (including real and imaginary numbers), and explain why the rectangular and polar forms of a given complex number represent the same number.	**Y**
N-CN.5	5. (+) Represent addition, subtraction, multiplication, and conjugation of complex numbers geometrically on the complex plane; use properties of this representation for computation. *For example,* $(-1 + \sqrt{3}i)^3 = 8$ *because* $(-1 + \sqrt{3}i)$ *has modulus 2 and argument 120°.*	**Y**
N-CN.6	6. (+) Calculate the distance between numbers in the complex plane as the modulus of the difference, and the midpoint of a segment as the average of the numbers at its endpoints.	**Y**
Use complex numbers in polynomial identities and equations.		
N-CN.7	7. Solve quadratic equations with real coefficients that have complex solutions.	**Y**
N-CN.8	8. (+) Extend polynomial identities to the complex numbers. *For example, rewrite* $x^2 + 4$ *as* $(x + 2i)(x - 2i)$.	**Y**
N-CN.9	9. (+) Know the Fundamental Theorem of Algebra; show that it is true for quadratic polynomials.	**Y**
Vector and Matrix Quantities [N-VM]		
Represent and model with vector quantities.		
N-VM.1	1. (+) Recognize vector quantities as having both magnitude and direction. Represent vector quantities by directed line segments, and use appropriate symbols for vectors and their magnitudes (e.g., $\mathbf{v}$, $\|\mathbf{v}\|$, $\|\|\mathbf{v}\|\|$, v).	**Y**
N-VM.2	2. (+) Find the components of a vector by subtracting the coordinates of an initial point from the coordinates of a terminal point.	**Y**
N-VM.3	3. (+) Solve problems involving velocity and other quantities that can be represented by vectors.	**Y**
Perform operations on vectors.		
N-VM.4	4. (+) Add and subtract vectors.	–
N-VM.4a	a. Add vectors end-to-end, component-wise, and by the parallelogram rule. Understand that the magnitude of a sum of two vectors is typically not the sum of the magnitudes.	**Y**

Code	Standard	Aligns
N-VM.4b	b. Given two vectors in magnitude and direction form, determine the magnitude and direction of their sum.	Y
N-VM.4c	c. Understand vector subtraction $\mathbf{v} - \mathbf{w}$ as $\mathbf{v} + (-\mathbf{w})$, where $-\mathbf{w}$ is the additive inverse of $\mathbf{w}$, with the same magnitude as $\mathbf{w}$ and pointing in the opposite direction. Represent vector subtraction graphically by connecting the tips in the appropriate order, and perform vector subtraction component-wise.	Y
N-VM.5	5. (+) Multiply a vector by a scalar.	
N-VM.5a	a. Represent scalar multiplication graphically by scaling vectors and possibly reversing their direction; perform scalar multiplication component-wise, e.g., as $c\,(vx, vy) = (cvx, cvy)$.	Y
N-VM.5b	b. Compute the magnitude of a scalar multiple $c\mathbf{v}$ using $\|\|c\mathbf{v}\|\| = \|c\|v$. Compute the direction of $c\mathbf{v}$ knowing that when $\|c\|v \neq 0$, the direction of $c\mathbf{v}$ is either along $\mathbf{v}$ (for $c > 0$) or against $\mathbf{v}$ (for $c < 0$).	Y
Perform operations on matrices and use matrices in applications.		
N-VM.6	6. (+) Use matrices to represent and manipulate data, e.g., to represent payoffs or incidence relationships in a network.	Y
N-VM.7	7. (+) Multiply matrices by scalars to produce new matrices, e.g., as when all of the payoffs in a game are doubled.	Y
N-VM.8	8. (+) Add, subtract, and multiply matrices of appropriate dimensions.	Y
N-VM.9	9. (+) Understand that, unlike multiplication of numbers, matrix multiplication for square matrices is not a commutative operation, but still satisfies the associative and distributive properties.	Y
N-VM.10	10. (+) Understand that the zero and identity matrices play a role in matrix addition and multiplication similar to the role of 0 and 1 in the real numbers. The determinant of a square matrix is nonzero if and only if the matrix has a multiplicative inverse.	Y
N-VM.11	11. (+) Multiply a vector (regarded as a matrix with one column) by a matrix of suitable dimensions to produce another vector. Work with matrices as transformations of vectors.	Y
N-VM.12	12. (+) Work with 2×2 matrices as transformations of the plane, and interpret the absolute value of the determinant in terms of area.	Y

* indicates Modeling standard (+) indicates standard beyond College and Career Ready

Code	Standard	Aligns
Algebra		
Seeing Structure in Expressions [A-SSE]		
Interpret the structure of expressions.		
A-SSE.1	1. Interpret expressions that represent a quantity in terms of its context.*	–
A-SSE.1a	a. Interpret parts of an expression, such as terms, factors, and coefficients.	**Y**
A-SSE.1b	b. Interpret complicated expressions by viewing one or more of their parts as a single entity. *For example, interpret $P(1 + r)^n$ as the product of P and a factor not depending on P.*	**Y**
A-SSE.2	2. Use the structure of an expression to identify ways to rewrite it. *For example, see $x^4 - y^4$ as $(x^2)^2 - (y^2)^2$, thus recognizing it as a difference of squares that can be factored as $(x^2 - y^2)(x^2 + y^2)$.*	**Y**
Write expressions in equivalent forms to solve problems.		
A-SSE.3	3. Choose and produce an equivalent form of an expression to reveal and explain properties of the quantity represented by the expression.*	–
A-SSE.3a	a. Factor a quadratic expression to reveal the zeros of the function it defines.	**Y**
A-SSE.3b	b. Complete the square in a quadratic expression to reveal the maximum or minimum value of the function it defines.	**Y**
A-SSE.3c	c. Use the properties of exponents to transform expressions for exponential functions. *For example, the expression 1.15^t can be rewritten as $(1.15^{1/12})^{12t} \approx 1.012^{12t}$ to reveal the approximate equivalent monthly interest rate if the annual rate is 15%.*	**Y**
A-SSE.4	4. Derive the formula for the sum of a finite geometric series (when the common ratio is not 1), and use the formula to solve problems. *For example, calculate mortgage payments.**	**Y**
Arithmetic with Polynomials and Rational Expressions [A-APR]		
Perform arithmetic operations on polynomials.		
A-APR.1	1. Understand that polynomials form a system analogous to the integers, namely, they are closed under the operations of addition, subtraction, and multiplication; add, subtract, and multiply polynomials.	**Y**

Code	Standard	Aligns
Understand the relationship between zeros and factors of polynomials.		
A-APR.2	2. Know and apply the Remainder Theorem: For a polynomial $p(x)$ and a number a, the remainder on division by $x - a$ is $p(a)$, so $p(a) = 0$ if and only if $(x - a)$ is a factor of $p(x)$.	Y
A-APR.3	3. Identify zeros of polynomials when suitable factorizations are available, and use the zeros to construct a rough graph of the function defined by the polynomial.	Y
Use polynomial identities to solve problems.		
A-APR.4	4. Prove polynomial identities and use them to describe numerical relationships. *For example, the polynomial identity $(x^2 + y^2)^2 = (x^2 - y^2)^2 + (2xy)^2$ can be used to generate Pythagorean triples.*	Y
A-APR.5	5. (+) Know and apply the Binomial Theorem for the expansion of $(x + y)^n$ in powers of x and y for a positive integer n, where x and y are any numbers, with coefficients determined for example by Pascal's Triangle.	Y
Rewrite rational expressions.		
A-APR.6	6. Rewrite simple rational expressions in different forms; write $a(x)/b(x)$ in the form $q(x) + r(x)/b(x)$, where $a(x)$, $b(x)$, $q(x)$, and $r(x)$ are polynomials with the degree of $r(x)$ less than the degree of $b(x)$, using inspection, long division, or, for the more complicated examples, a computer algebra system.	Y
A-APR.7	7. (+) Understand that rational expressions form a system analogous to the rational numbers, closed under addition, subtraction, multiplication, and division by a nonzero rational expression; add, subtract, multiply, and divide rational expressions.	Y
Creating Equations [A-CED]		
Create equations that describe numbers or relationships.		
A-CED.1	1. Create equations and inequalities in one variable and use them to solve problems. *Include equations arising from linear and quadratic functions, and simple rational and exponential functions.*	Y

(Continued)

Code	Standard	Aligns
A-CED.2	2. Create equations in two or more variables to represent relationships between quantities; graph equations on coordinate axes with labels and scales.	**Y**
A-CED.3	3. Represent constraints by equations or inequalities, and by systems of equations and/or inequalities, and interpret solutions as viable or non-viable options in a modeling context. *For example, represent inequalities describing nutritional and cost constraints on combinations of different foods.*	**Y**
A-CED.4	4. Rearrange formulas to highlight a quantity of interest, using the same reasoning as in solving equations. *For example, rearrange Ohm's law $V = IR$ to highlight resistance R.*	**Y**
Reasoning with Equations and Inequalities [A-REI]		
Understand solving equations as a process of reasoning and explain the reasoning.		
A-REI.1	1. Explain each step in solving a simple equation as following from the equality of numbers asserted at the previous step, starting from the assumption that the original equation has a solution. Construct a viable argument to justify a solution method.	**Y**
A-REI.2	2. Solve simple rational and radical equations in one variable, and give examples showing how extraneous solutions may arise.	**Y**
Solve equations and inequalities in one variable.		
A-REI.3	3. Solve linear equations and inequalities in one variable, including equations with coefficients represented by letters.	**Y**
A-REI.4	4. Solve quadratic equations in one variable.	–
A-REI.4a	a. Use the method of completing the square to transform any quadratic equation in x into an equation of the form $(x - p)^2 = q$ that has the same solutions. Derive the quadratic formula from this form.	**Y**

Code	Standard	Aligns
A-REI.4b	b. Solve quadratic equations by inspection (e.g., for $x^2 = 49$), taking square roots, completing the square, the quadratic formula, and factoring, as appropriate to the initial form of the equation. Recognize when the quadratic formula gives complex solutions and write them as $a \pm bi$ for real numbers a and b.	**Y**
Solve systems of equations.		
A-REI.5	5. Prove that, given a system of two equations in two variables, replacing one equation by the sum of that equation and a multiple of the other produces a system with the same solutions.	**Y**
A-REI.6	6. Solve systems of linear equations exactly and approximately (e.g., with graphs), focusing on pairs of linear equations in two variables.	**Y**
A-REI.7	7. Solve a simple system consisting of a linear equation and a quadratic equation in two variables algebraically and graphically. *For example, find the points of intersection between the line $y = -3x$ and the circle $x^2 + y^2 = 3$.*	**Y**
A-REI.8	8. (+) Represent a system of linear equations as a single matrix equation in a vector variable.	**Y**
A-REI.9	9. (+) Find the inverse of a matrix if it exists and use it to solve systems of linear equations (using technology for matrices of dimension 3 × 3 or greater).	**P**
Represent and solve equations and inequalities graphically.		
A-REI.10	10. Understand that the graph of an equation in two variables is the set of all its solutions plotted in the coordinate plane, often forming a curve (which could be a line).	**Y**
A-REI.11	11. Explain why the x-coordinates of the points where the graphs of the equations $y = f(x)$ and $y = g(x)$ intersect are the solutions of the equation $f(x) = g(x)$; find the solutions approximately, e.g., using technology to graph the functions, make tables of values, or find successive approximations. Include cases where $f(x)$ and/or $g(x)$ are linear, polynomial, rational, absolute value, exponential, and logarithmic functions.*	**P**

(Continued)

Code	Standard	Aligns
A-REI.12	12. Graph the solutions to a linear inequality in two variables as a half-plane (excluding the boundary in the case of a strict inequality), and graph the solution set to a system of linear inequalities in two variables as the intersection of the corresponding half-planes.	**Y**

* indicates Modeling standard (+) indicates standard beyond College and Career Ready

Code	Standard	Aligns
Functions		
Interpreting Functions [F-IF]		
Understand the concept of a function and use function notation.		
F-IF.1	1. Understand that a function from one set (called the domain) to another set (called the range) assigns to each element of the domain exactly one element of the range. If f is a function and x is an element of its domain, then $f(x)$ denotes the output of f corresponding to the input x. The graph of f is the graph of the equation $y = f(x)$.	**Y**
F-IF.2	2. Use function notation, evaluate functions for inputs in their domains, and interpret statements that use function notation in terms of a context.	**Y**
F-IF.3	3. Recognize that sequences are functions, sometimes defined recursively, whose domain is a subset of the integers. For example, the Fibonacci sequence is defined recursively by $f(0) = f(1) = 1$, $f(n + 1) = f(n) + f(n - 1)$ for $n \geq 1$.	**Y**
Interpret functions that arise in applications in terms of the context.		
F-IF.4	4. For a function that models a relationship between two quantities, interpret key features of graphs and tables in terms of the quantities, and sketch graphs showing key features given a verbal description of the relationship. *Key features include: intercepts; intervals where the function is increasing, decreasing, positive, or negative; relative maximums and minimums; symmetries; end behavior; and periodicity**	**Y**
F-IF.5	5. Relate the domain of a function to its graph and, where applicable, to the quantitative relationship it describes. *For example, if the function h(n) gives the number of person-hours it takes to assemble n engines in a factory, then the positive integers would be an appropriate domain for the function.**	**Y**
F-IF.6	6. Calculate and interpret the average rate of change of a function (presented symbolically or as a table) over a specified interval. Estimate the rate of change from a graph.*	**Y**
Analyze functions using different representations.		
F-IF.7	7. Graph functions expressed symbolically and show key features of the graph, by hand in simple cases and using technology for more complicated cases.*	

(Continued)

Code	Standard	Aligns
F-IF.7a	a. Graph linear and quadratic functions and show intercepts, maxima, and minima.*	
F-IF.7b	b. Graph square root, cube root, and piecewise-defined functions, including step functions and absolute value functions.	
F-IF.7c	c. Graph polynomial functions, identifying zeros when suitable factorizations are available, and showing end behavior.	**P**
F-IF.7d	d. (+) Graph rational functions, identifying zeros and asymptotes when suitable factorizations are available, and showing end behavior.*	
F-IF.7e	e. Graph exponential and logarithmic functions, showing intercepts and end behavior, and trigonometric functions, showing period, midline, and amplitude.*	
F-IF.8	8. Write a function defined by an expression in different but equivalent forms to reveal and explain different properties of the function.	–
F-IF.8a	a. Use the process of factoring and completing the square in a quadratic function to show zeros, extreme values, and symmetry of the graph, and interpret these in terms of a context.	**Y**
F-IF.8b	b. Use the properties of exponents to interpret expressions for exponential functions. *For example, identify percent rate of change in functions such as* $y = (1.02)^t$, $y = (0.97)^t$, $y = (1.01)^{12t}$, *and* $y = (1.2)^{t/10}$, *and classify them as representing exponential growth or decay.*	**Y**
F-IF.9	9. Compare properties of two functions each represented in a different way (algebraically, graphically, numerically in tables, or by verbal descriptions). *For example, given a graph of one quadratic function and an algebraic expression for another, say which has the larger maximum.*	**Y**
Building Functions [F-BF]		
Build a function that models a relationship between two quantities.		
F-BF.1	1. Write a function that describes a relationship between two quantities.*	–
F-BF.1a	a. Determine an explicit expression, a recursive process, or steps for calculation from a context.*	**Y**
F-BF.1b	b. Combine standard function types using arithmetic operations. *For example, build a function that models the temperature of a cooling body by adding a constant function to a decaying exponential, and relate these functions to the model.**	**Y**

Code	Standard	Aligns
F-BF.1c	c. (+) Compose *functions. For example, if T(y) is the temperature in the atmosphere as a function of height, and h(t) is the height of a weather balloon as a function of time, then T(h(t)) is the temperature at the location of the weather balloon as a function of time.**	**Y**
F-BF.2	2. Write arithmetic and geometric sequences both recursively and with an explicit formula, use them to model situations, and translate between the two forms.*	**Y**
Build new functions from existing functions.		
F-BF.3	3. Identify the effect on the graph of replacing $f(x)$ by $f(x) + k$, $k\,f(x)$, $f(kx)$, and $f(x + k)$ for specific values of k (both positive and negative); find the value of k given the graphs. Experiment with cases and illustrate an explanation of the effects on the graph using technology. *Include recognizing even and odd functions from their graphs and algebraic expressions for them.*	**Y**
F-BF.4	4. Find inverse functions.	–
F-BF.4a	a. Solve an equation of the form $f(x) = c$ for a simple function f that has an inverse and write an expression for the inverse. *For example,* $f(x) = 2x^3$ *or* $f(x) = (x + 1)/(x – 1)$ *for* $x \neq 1$.	**Y**
F-BF.4b	b. (+) Verify by composition that one function is the inverse of another.	**Y**
F-BF.4c	c. (+) Read values of an inverse function from a graph or a table, given that the function has an inverse.	**Y**
F-BF.4d	d. (+) Produce an invertible function from a non-invertible function by restricting the domain.	**Y**
F-BF.5	5. (+) Understand the inverse relationship between exponents and logarithms and use this relationship to solve problems involving logarithms and exponents.	**Y**
Linear, Quadratic, and Exponential Models [F-LE]		
Construct and compare linear, quadratic, and exponential models and solve problems.		
F-LE.1	1. Distinguish between situations that can be modeled with linear functions and with exponential functions.*	–
F-LE.1a	a. Prove that linear functions grow by equal differences over equal intervals, and that exponential functions grow by equal factors over equal intervals.*	**Y**
F-LE.1b	b. Recognize situations in which one quantity changes at a constant rate per unit interval relative to another.*	**Y**

(Continued)

Code	Standard	Aligns
F-LE.1c	c. Recognize situations in which a quantity grows or decays by a constant percent rate per unit interval relative to another.*	**Y**
F-LE.2	2. Construct linear and exponential functions, including arithmetic and geometric sequences, given a graph, a description of a relationship, or two input-output pairs (include reading these from a table).*	**Y**
F-LE.3	3. Observe using graphs and tables that a quantity increasing exponentially eventually exceeds a quantity increasing linearly, quadratically, or (more generally) as a polynomial function.*	**Y**
F-LE.4	4. For exponential models, express as a logarithm the solution to $ab^{ct} = d$ where a, c, and d are numbers and the base b is 2, 10, or e; evaluate the logarithm using technology.*	**P**
Interpret expressions for functions in terms of the situation they model.		
F-LE.5	5. Interpret the parameters in a linear or exponential function in terms of a context.*	**Y**
Trigonometric Functions [F-TF]		
Extend the domain of trigonometric functions using the unit circle.		
F-TF.1	1. Understand radian measure of an angle as the length of the arc on the unit circle subtended by the angle.	**Y**
F-TF.2	2. Explain how the unit circle in the coordinate plane enables the extension of trigonometric functions to all real numbers, interpreted as radian measures of angles traversed counterclockwise around the unit circle.	**Y**
F-TF.3	3. (+) Use special triangles to determine geometrically the values of sine, cosine, tangent for $\pi/3$, $\pi/4$ and $\pi/6$, and use the unit circle to express the values of sine, cosine, and tangent for $\pi - x$, $\pi + x$, and $2\pi - x$ in terms of their values for x, where x is any real number.	**Y**
F-TF.4	4. (+) Use the unit circle to explain symmetry (odd and even) and periodicity of trigonometric functions.	**Y**
Model periodic phenomena with trigonometric functions.		
F-TF.5	5. Choose trigonometric functions to model periodic phenomena with specified amplitude, frequency, and midline.*	**Y**
F-TF.6	6. (+) Understand that restricting a trigonometric function to a domain on which it is always increasing or always decreasing allows its inverse to be constructed.	**Y**

Code	Standard	Aligns
F-TF.7	7. (+) Use inverse functions to solve trigonometric equations that arise in modeling contexts; evaluate the solutions using technology, and interpret them in terms of the context.*	**P**
Prove and apply trigonometric identities.		
F-TF.8	8. Prove the Pythagorean identity $\sin^2(\theta) + \cos^2(\theta) = 1$ and use it to find $\sin(\theta)$, $\cos(\theta)$, or $\tan(\theta)$ given $\sin(\theta)$, $\cos(\theta)$, or $\tan(\theta)$ and the quadrant.	**Y**
F-TF.9	9. (+) Prove the addition and subtraction formulas for sine, cosine, and tangent and use them to solve problems.	**Y**

* indicates Modeling standard (+) indicates standard beyond College and Career Ready

Code	Standard	Aligns
Geometry		
Congruence [G-CO]		
Experiment with transformations in the plane.		
G-CO.1	1. Know precise definitions of angle, circle, perpendicular line, parallel line, and line segment, based on the undefined notions of point, line, distance along a line, and distance around a circular arc.	**Y**
G-CO.2	2. Represent transformations in the plane using, e.g., transparencies and geometry software; describe transformations as functions that take points in the plane as inputs and give other points as outputs. Compare transformations that preserve distance and angle to those that do not (e.g., translation versus horizontal stretch).	**P**
G-CO.3	3. Given a rectangle, parallelogram, trapezoid, or regular polygon, describe the rotations and reflections that carry it onto itself.	**Y**
G-CO.4	4. Develop definitions of rotations, reflections, and translations in terms of angles, circles, perpendicular lines, parallel lines, and line segments.	**Y**
G-CO.5	5. Given a geometric figure and a rotation, reflection, or translation, draw the transformed figure using, e.g., graph paper, tracing paper, or geometry software. Specify a sequence of transformations that will carry a given figure onto another.	**P**
Understand congruence in terms of rigid motions.		
G-CO.6	6. Use geometric descriptions of rigid motions to transform figures and to predict the effect of a given rigid motion on a given figure; given two figures, use the definition of congruence in terms of rigid motions to decide if they are congruent.	**Y**
G-CO.7	7. Use the definition of congruence in terms of rigid motions to show that two triangles are congruent if and only if corresponding pairs of sides and corresponding pairs of angles are congruent.	**Y**
G-CO.8	8. Explain how the criteria for triangle congruence (ASA, SAS, and SSS) follow from the definition of congruence in terms of rigid motions.	**Y**
Prove geometric theorems.		
G-CO.9	9. Prove theorems about lines and angles. *Theorems include: vertical angles are congruent; when a transversal crosses parallel lines, alternate interior angles are congruent and corresponding angles are congruent; points on a perpendicular bisector of a line segment are exactly those equidistant from the segment's endpoints.*	**Y**

Code	Standard	Aligns
G-CO.10	10. Prove theorems about triangles. *Theorems include: measures of interior angles of a triangle sum to 180°; base angles of isosceles triangles are congruent; the segment joining midpoints of two sides of a triangle is parallel to the third side and half the length; the medians of a triangle meet at a point.*	**Y**
G-CO.11	11. Prove theorems about parallelograms. *Theorems include: opposite sides are congruent, opposite angles are congruent, the diagonals of a parallelogram bisect each other, and conversely, rectangles are parallelograms with congruent diagonals.*	**Y**
Make geometric constructions.		
G-CO.12	12. Make formal geometric constructions with a variety of tools and methods (compass and straightedge, string, reflective devices, paper folding, dynamic geometric software, etc.). *Copying a segment; copying an angle; bisecting a segment; bisecting an angle; constructing perpendicular lines, including the perpendicular bisector of a line segment; and constructing a line parallel to a given line through a point not on the line.*	**P**
G-CO.13	13. Construct an equilateral triangle, a square, and a regular hexagon inscribed in a circle.	**Y**
Similarity, Right Triangles, and Trigonometry [G-SRT]		
Understand similarity in terms of similarity transformations.		
G-SRT.1	1. Verify experimentally the properties of dilations given by a center and a scale factor:	-
G-SRT.1a	a. A dilation takes a line not passing through the center of the dilation to a parallel line, and leaves a line passing through the center unchanged.	**Y**
G-SRT.1b	b. The dilation of a line segment is longer or shorter in the ratio given by the scale factor.	**Y**
G-SRT.2	2. Given two figures, use the definition of similarity in terms of similarity transformations to decide if they are similar; explain using similarity transformations the meaning of similarity for triangles as the equality of all corresponding pairs of angles and the proportionality of all corresponding pairs of sides.	**Y**
G-SRT.3	3. Use the properties of similarity transformations to establish the Angle-Angle (AA) criterion for two triangles to be similar.	**Y**

(Continued)

Code	Standard	Aligns
Prove theorems involving similarity.		
G-SRT.4	4. Prove theorems about triangles. *Theorems include: a line parallel to one side of a triangle divides the other two proportionally, and conversely; the Pythagorean Theorem proved using triangle similarity.*	**Y**
G-SRT.5	5. Use congruence and similarity criteria for triangles to solve problems and to prove relationships in geometric figures.	**Y**
Define trigonometric ratios and solve problems involving right triangles.		
G-SRT.6	6. Understand that by similarity, side ratios in right triangles are properties of the angles in the triangle, leading to definitions of trigonometric ratios for acute angles.	**Y**
G-SRT.7	7. Explain and use the relationship between the sine and cosine of complementary angles.	**Y**
G-SRT.8	8. Use trigonometric ratios and the Pythagorean Theorem to solve right triangles in applied problems.*	**Y**
Apply trigonometry to general triangles.		
G-SRT.9	9. (+) Derive the formula $A = \frac{1}{2}ab \sin(C)$ for the area of a triangle by drawing an auxiliary line from a vertex perpendicular to the opposite side.	**Y**
G-SRT.10	10. (+) Prove the Laws of Sines and Cosines and use them to solve problems.	**Y**
G-SRT.11	11. (+) Understand and apply the law of sines and the law of cosines to find unknown measurements in right and non-right triangles (e.g., surveying problems, resultant forces).	**Y**
Circles [G-C]		
Understand and apply theorems about circles.		
G-C.1	1. Prove that all circles are similar.	**Y**
G-C.2	2. Identify and describe relationships among inscribed angles, radii, and chords. *Include the relationship between central, inscribed, and circumscribed angles; inscribed angles on a diameter are right angles; the radius of a circle is perpendicular to the tangent where the radius intersects the circle.*	**Y**
G-C.3	3. Construct the inscribed and circumscribed circles of a triangle, and prove properties of angles for a quadrilateral inscribed in a circle.	**Y**
G-C.4	4. (+) Construct a tangent line from a point outside a given circle to the circle.	**Y**

Code	Standard	Aligns
Find arc lengths and areas of sectors of circles.		
G-C.5	5. Derive using similarity the fact that the length of the arc intercepted by an angle is proportional to the radius, and define the radian measure of the angle as the constant of proportionality; derive the formula for the area of a sector.	**Y**
Expressing Geometric Properties with Equations [G-GPE]		
Translate between the geometric description and the equation for a conic section.		
G-GPE.1	1. Derive the equation of a circle of given center and radius using the Pythagorean Theorem; complete the square to find the center and radius of a circle given by an equation.	**Y**
G-GPE.2	2. Derive the equation of a parabola given a focus and directrix.	**Y**
G-GPE.3	3. (+) Derive the equations of ellipses and hyperbolas given the foci, using the fact that the sum or difference of distances from the foci is constant.	**Y**
Use coordinates to prove simple geometric theorems algebraically.		
G-GPE.4	4. Use coordinates to prove simple geometric theorems algebraically. For example, prove or disprove that a figure defined by four given points in the coordinate plane is a rectangle; prove or disprove that the point $(1,\sqrt{3})$ lies on the circle centered at the origin and containing the point (0, 2).	**Y**
G-GPE.5	5. Prove the slope criteria for parallel and perpendicular lines and use them to solve geometric problems (e.g., find the equation of a line parallel or perpendicular to a given line that passes through a given point).	**Y**
G-GPE.6	6. Find the point on a directed line segment between two given points that partitions the segment in a given ratio.	**Y**
G-GPE.7	7. Use coordinates to compute perimeters of polygons and areas of triangles and rectangles, e.g., using the distance formula.*	**Y**
Geometric Measurement and Dimension [G-GMD]		
Explain volume formulas and use them to solve problems.		
G-GMD.1	1. Give an informal argument for the formulas for the circumference of a circle, area of a circle, volume of a cylinder, pyramid, and cone. *Use dissection arguments, Cavalieri's principle, and informal limit arguments.*	**Y**

(Continued)

Code	Standard	Aligns
G-GMD.2	2. (+) Give an informal argument using Cavalieri's principle for the formulas for the volume of a sphere and other solid figures.	**Y**
G-GMD.3	3. Use volume formulas for cylinders, pyramids, cones, and spheres to solve problems.*	**Y**
Visualize relationships between two-dimensional and three-dimensional objects.		
G-GMD.4	4. Identify the shapes of two-dimensional cross-sections of three-dimensional objects, and identify three-dimensional objects generated by rotations of two-dimensional objects.	**Y**
Modeling with Geometry [G-MG]		
Apply geometric concepts in modeling situations.		
G-MG.1	1. Use geometric shapes, their measures, and their properties to describe objects (e.g., modeling a tree trunk or a human torso as a cylinder).*	**Y**
G-MG.2	2. Apply concepts of density based on area and volume in modeling situations (e.g., persons per square mile, BTUs per cubic foot.*	**Y**
G-MG.3	3. Apply geometric methods to solve design problems (e.g., designing an object or structure to satisfy physical constraints or minimize cost; working with typographic grid systems based on ratios).*	**Y**

* indicates Modeling standard (+) indicates standard beyond College and Career Ready

Code	Standard	Aligns
Statistics and Probability		
Interpreting Categorical and Quantitative Data [S-ID]		
Summarize, represent, and interpret data on a single count or measurement variable.		
S-ID.1	1. Represent data with plots on the real number line (dot plots, histograms, and box plots).*	**Y**
S-ID.2	2. Use statistics appropriate to the shape of the data distribution to compare center (median, mean) and spread (interquartile range, standard deviation) of two or more different data sets.*	**Y**
S-ID.3	3. Interpret differences in shape, center, and spread in the context of the data sets, accounting for possible effects of extreme data points (outliers).*	**Y**
S-ID.4	4. Use the mean and standard deviation of a data set to fit it to a normal distribution and to estimate population percentages. Recognize that there are data sets for which such a procedure is not appropriate. Use calculators, spreadsheets, and tables to estimate areas under the normal curve.	**Y**
Summarize, represent, and interpret data on two categorical and quantitative variables.		
S-ID.5	5. Summarize categorical data for two categories in two-way frequency tables. Interpret relative frequencies in the context of the data (including joint, marginal, and conditional relative frequencies). Recognize possible associations and trends in the data.*	**Y**
S-ID.6	6. Represent data on two quantitative variables on a scatter plot, and describe how the variables are related.*	**-**
S-ID.6a	a. Fit a function to the data; use functions fitted to data to solve problems in the context of the data. Use given functions or choose a function suggested by the context. Emphasize linear, quadratic, and exponential models.*	**Y**
S-ID.6b	b. Informally assess the fit of a function by plotting and analyzing residuals.*	**Y**
S-ID.6c	c. Fit a linear function for a scatter plot that suggests a linear association.*	**Y**
Interpret linear models.		
S-ID.7	7. Interpret the slope (rate of change) and the intercept (constant term) of a linear model in the context of the data.*	**Y**

(Continued)

Code	Standard	Aligns
S-ID.8	8. Compute (using technology) and interpret the correlation coefficient of a linear fit.*	Y
S-ID.9	9. Distinguish between correlation and causation.*	Y

Making Inferences and Justifying Conclusions [S-IC]

Understand and evaluate random processes underlying statistical experiments.

S-IC.1	1. Understand statistics as a process for making inferences to be made about population parameters based on a random sample from that population.*	Y
S-IC.2	2. Decide if a specified model is consistent with results from a given data-generating process, e.g., using simulation. *For example, a model says a spinning coin falls heads up with probability 0.5. Would a result of 5 tails in a row cause you to question the model?**	Y

Make inferences and justify conclusions from sample surveys, experiments, and observational studies.

S-IC.3	3. Recognize the purposes of and differences among sample surveys, experiments, and observational studies; explain how randomization relates to each.*	Y
S-IC.4	4. Use data from a sample survey to estimate a population mean or proportion; develop a margin of error through the use of simulation models for random sampling.*	Y
S-IC.5	5. Use data from a randomized experiment to compare two treatments; use simulations to decide if differences between parameters are significant.*	Y
S-IC.6	6. Evaluate reports based on data.*	Y

Conditional Probability and the Rules of Probability [S-CP]

Understand independence and conditional probability and use them to interpret data.

S-CP.1	1. Describe events as subsets of a sample space (the set of outcomes) using characteristics (or categories) of the outcomes, or as unions, intersections, or complements of other events ("or", "and", "not").*	Y
S-CP.2	2. Understand that two events A and B are independent if the probability of A and B occurring together is the product of their probabilities, and use this characterization to determine if they are independent.*	Y

Code	Standard	Aligns
S-CP.3	3. Understand the conditional probability of *A* given *B* as *P*(*A* and *B*)/*P*(*B*), and interpret independence of *A* and *B* as saying that the conditional probability of *A* given *B* is the same as the probability of *A*, and the conditional probability of *B* given *A* is the same as the probability of *B*.*	**Y**
S-CP.4	4. Construct and interpret two-way frequency tables of data when two categories are associated with each object being classified. Use the two-way table as a sample space to decide if events are independent and to approximate conditional probabilities. *For example, collect data from a random sample of students in your school on their favorite subject among math, science, and English. Estimate the probability that a randomly selected student from your school will favor science given that the student is in tenth grade. Do the same for other subjects and compare the results.**	**Y**
S-CP.5	5. Recognize and explain the concepts of conditional probability and independence in everyday language and everyday situations. *For example, compare the chance of having lung cancer if you are a smoker with the chance of being a smoker if you have lung cancer.**	**Y**
Use the rules of probability to compute probabilities of compound events in a uniform probability model.		
S-CP.6	6. Find the conditional probability of *A* given *B* as the fraction of *B*'s outcomes that also belong to *A*, and interpret the answer in terms of the model.*	**Y**
S-CP.7	7. Apply the Addition Rule, $P(A \text{ or } B) = P(A) + P(B) - P(A \text{ and } B)$, and interpret the answer in terms of the model.*	**Y**
S-CP.8	8. (+) Apply the general Multiplication Rule in a uniform probability model, $P(A \text{ and } B) = P(A)P(B\|A) = P(B)P(A\|B)$, and interpret the answer in terms of the model.*	**Y**
S-CP.9	9. (+) Use permutations and combinations to compute probabilities of compound events and solve problems.*	**Y**
Using Probability to Make Decisions [S-MD]		
Calculate expected values and use them to solve problems.		
S-MD.1	1. (+) Define a random variable for a quantity of interest by assigning a numerical value to each event in a sample space; graph the corresponding probability distribution using the same graphical displays as for data distributions.*	**Y**

(Continued)

Code	Standard	Aligns
S-MD.2	2. (+) Calculate the expected value of a random variable; interpret it as the mean of the probability distribution.*	**Y**
S-MD.3	3. (+) Develop a probability distribution for a random variable defined for a sample space in which theoretical probabilities can be calculated; find the expected value. *For example, find the theoretical probability distribution for the number of correct answers obtained by guessing on all five questions of a multiple-choice test where each question has four choices, and find the expected grade under various grading schemes.**	**Y**
S-MD.4	4. (+) Develop a probability distribution for a random variable defined for a sample space in which probabilities are assigned empirically; find the expected value. *For example, find a current data distribution on the number of TV sets per household in the United States, and calculate the expected number of sets per household. How many TV sets would you expect to find in 100 randomly selected households?**	**Y**
Use probability to evaluate outcomes of decisions.		
S-MD.5	5. (+) Weigh the possible outcomes of a decision by assigning probabilities to payoff values and finding expected values.	**-**
S-MD.5a	a. (+) Find the expected payoff for a game of chance. For example, find the expected winnings from a state lottery ticket or a game at a fast-food restaurant.*	**Y**
S-MD.5b	b. (+) Evaluate and compare strategies on the basis of expected values. For example, compare a high-deductible versus a low-deductible automobile insurance policy using various, but reasonable, chances of having a minor or a major accident.*	**Y**
S-MD.6	6. (+) Use probabilities to make fair decisions (e.g., drawing by lots, using a random number generator).*	**Y**
S-MD.7	7. (+) Analyze decisions and strategies using probability concepts (e.g., product testing, medical testing, pulling a hockey goalie at the end of a game).*	**Y**

*indicates Modeling standard (+) indicates standard beyond College and Career Ready

Science Standards

The following table shows only the standards in this strand that align to the ACT Science.

Code	Standard	Aligns
R.CCR.7	**Integrate and evaluate content presented in diverse formats and media, including visually and quantitatively, as well as in words.**	***P***
RI.11–12.7	Integrate and evaluate multiple sources of information presented in different media or formats (e.g., visually, quantitatively) as well as in words in order to address a question or solve a problem.	***P***
R.CCR.8	**Delineate and evaluate the argument and specific claims in a text, including the validity of the reasoning as well as the relevance and sufficiency of the evidence.**	***Y***
R.CCR.9	**Analyze how two or more texts address similar themes or topics in order to build knowledge or to compare the approaches the authors take.**	***Y***

The following table shows all standards in the RST strand, some of which align with ACT Science and others with ACT Reading.

Code	Standard	Aligns
Key Ideas and Details		
RST.11–12.1	Cite specific textual evidence to support analysis of science and technical texts, attending to important distinctions the author makes and to any gaps or inconsistencies in the account.	**PR**
RST.11–12.2	Determine the central ideas or conclusions of a text; summarize complex concepts, processes or information presented in a text by paraphrasing them in simpler but still accurate terms.	**YR**
RST.11–12.3	Follow precisely a complex multistep procedure when carrying out experiments, taking measurements, or performing technical tasks; analyze the specific results based on explanations in the text.	**YS**
Craft and Structure		
RST.11–12.4	Determine the meaning of symbols, key terms, and other domain-specific words and phrases as they are used in a specific scientific or technical context relevant to grades 11–12 texts and topics.	**PR**
RST.11–12.5	Analyze how the text structures information or ideas into categories or hierarchies, demonstrating understanding of the information or ideas.	**YR**
RST.11–12.6	Analyze the author's purpose in providing an explanation, describing a procedure, or discussing an experiment in a text, identifying important issues that remain unresolved.	**YR**
Integration of Knowledge and Ideas		
RST.11–12.7	Integrate and evaluate multiple sources of information presented in diverse formats and media (e.g., quantitative data, video, multimedia) in order to address a question or solve a problem.	**PS**
RST.11–12.8	Evaluate the hypotheses, data, analysis, and conclusions in a science or technical text, verifying the data when possible and corroborating or challenging conclusions with other sources of information.	**YS**
RST.11–12.9	Synthesize information from a range of sources (e.g. texts, experiments, simulations) into a coherent understanding of a process, phenomenon, or concept, resolving conflicting information when possible.	**YS**
Range of Reading and Level of Text Complexity		
RST.11–12.10	By the end of grade 12, read and comprehend science/technical texts in the grades 11-CCR text complexity band independently and proficiently.	**YRS**

Additional Resources

A-List

Main website: www.alisteducation.com
Bookstore: www.alisteducation.com/bookstore

The ACT

Main websites:
www.act.org
www.actstudent.org

The Alignment of Common Core and ACT's College and Career Readiness System, ACT, Inc., June 2010. [No longer posted on act.org]

The Official ACT Prep Guide 2016–2017, Wiley, 2016.
An ACT, Inc., book that contains three full-length practice ACTs

The Real ACT Prep Guide, Third 3rd Edition, Peterson's, 2011.
An earlier version of the book that contains five full-length practice ACTs. There is some overlap between the material in these tests and the material in the tests in The Official ACT Prep Guide. Note Wiley produced a 2016 printing of the same book. Be sure to pay attention to the titles when ordering books.

Preparing for the ACT 2015–2016, free booklet containing a full practice test:
www.act.org/content/dam/act/unsecured/documents/Preparing-for-the-ACT.pdf

Practice test booklet order form:
www.act.org/content/dam/act/unsecured/documents/ACT-SampleTestBookOrderForm.pdf

ACT College and Career Readiness Standards:
www.act.org/content/act/en/education-and-career-planning/CollegeandCareerReadinessStandards.html

The Common Core State Standards

Main website:
www.corestandards.org

The Standards (available to read on the web or as pdf downloads):
www.corestandards.org/the-standards

For Product Safety Concerns and Information please contact our EU
representative GPSR@taylorandfrancis.com
Taylor & Francis Verlag GmbH, Kaufingerstraße 24, 80331 München, Germany

www.ingramcontent.com/pod-product-compliance
Ingram Content Group UK Ltd.
Pitfield, Milton Keynes, MK11 3LW, UK
UKHW052222190926
475702UK00006B/1152
* 9 7 8 1 1 3 8 6 9 2 2 0 6 *